THE GREAT OUTDOORS BOOK OF

SEASHORE LIFE

by CRICKET HARRIS

GREAT OUTDOORS PUBLISHING CO.
4747 TWENTY-EIGHTH STREET NORTH
ST. PETERSBURG, FLORIDA 33714

About the Author

CRICKET HARRIS

DEDICATION

In memory of my husband "Cappy" who was always my best friend.

"Putting this book together has been one of the most satisfying things I have ever done," so spoke the author of this book, Mrs. Barnett Harris, better known as "Cricket" to all whom she has met.

When asked how long she has been interested in natural history Cricket will tell you, "Since I first sat on an ant hill and found such small things could hurt so much.

Asked about her education in preparation for this work she will tell you that she failed her final biology test in college by putting head down on a tray containing an "opened" anaesthises toad and weeping bitterly.

At the University of Chicago and Northwestern University she was allowed to sit and listen in many classes, and loaned some of the finest books of that time from personal libraries belonging to renouned teachers.

Together she and Captain Harris made the first big set of educational nature study films for the Society of Visual Education. She studied and worked with him on the "Mercy Bullet," the first humane means of capturing wild animals, and accompanied him to Africa on a scientific collecting expedition in 1928, and to prove his "Mercy Bullet." They went into the bush country to capture animals for the Brookfield Zoo, but the depression cut their work short and did horrible financial things to them, but they finally made it home — bringing back only one small creature, an infant daughter, but with a deeper friendship and an unswerving faith in one another they went to work to organize their material and knowledge and brought this book into being.

Cricket taught natural science in the Junior High School, in Evanston, Illinois, writing her own text book from day to day. But all that goes back to her childhood-summers spent in wild stretches along the Ohio River, and winters spent in Chicago, where a kindly elderly gentleman in the Academy of Sciences, in Lincoln Park, taught her to collect and preserve pond and field insects and weeds. Gentle people, including Mr. Davies, at the field Museum — now the Chicago Museum of Natural History — encouraged her and let her watch the intricate work that goes into creating a museum.

So with a deep apreciation of all that has been done for her along the way, she has compiled this book on Seashore Life.

CONTENTS

Foreword

WHEN YOU step into the edge of the sea, you are stepping out of your everyday world into a fairy-land filled with beauty and ugliness, and a strangeness almost fantastically unbelievable. The shore-line you walk today was not here yesterday, and will not be here to-morrow — it is ever changing. Forms that are washed ashore today may not be seen again for a long time, so gather what is there; of the dead fill your basket — of the living take only what you believe is right. Every thing in the ocean is there for a purpose, for Nature is all-wise, and the balance, even in a small area, must not be upset. An overturned stone or shell should be carefully replaced to spare the creatures living upon it or beneath it.

Many things are very small, but they are so numerous that you see them when massed. These creatures are referred to as minutens. They are the first animals — the protozoans — whose thousands of species are to be found anywhere and everywhere in the world.

Wind and water transport the protozoans, scattering them over the world, but should they land where conditions are unfavorable, these minutens enclose themselves in a tough coating remaining dormant for long periods of time.

Some protozoans are merely bits of living jelly, others are unbelievably strange and lovely.

One well known of these tiny creatures is Gymnodinium brevis, the protozoan that causes the dread "red tide" which turns great patches of sea water a dull brown-red. All creatures that breathe this water die of asphyxiation and suffocation. "Red tides" occur the world over in all warm seas.

Gymnodium brevis, however, does not cause the red color in the Red Sea — this is done by a small plant known as an alga.

All animals shown in this book are invertebrates—animals without backbones.

Acknowledgments

This book is for the layman who picks up many of these creatures along our shores and would like to know what he has found. It was made possible by the people from ancient times up to today who have identified backboneless creatures and observed their habits. To all these, and the many friends who have lent their prize specimens to help out, I am most grateful. Prof. Clyde T. Reed of the University of Tampa, Fla. has been an invaluable friend.

SEASHORE LIFE

Those who walk the beaches, whether hunting shells or just for pleasure are going to find many small living creatures stranded after a high tide that are unfamiliar and so are a bit frightening. This book is to answer the eternal question, "What is it?" — "Will it hurt me?"

THE text is compiled on a definite pattern, but only fundamentals are given. The book is laid out in such a way that it may be usable to the average person, and as a text from grade schools through college.

The invertebrates, or creatures without backbones, illustrated herein are described in the following pages as to common names, scientific names with authorities given, size and color, food and locality, and known habits. Sometimes the information is scant for less is known about the seashore animals than about any other of earth's creatures.

The subject matter begins with the minutens —the foraminifera — tiny creatures whose remains are in the chalk used at school. Following these are the sponges — the first many-celled animals, hatched from eggs and growing like plants. Some are individuals but most are colonial.

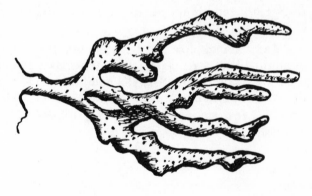

This is one of the finger sponges. Its skeleton cannot be used commercially.

A sponge hatches from an egg and grows like a plant. As an infant it is a delicate creature that has hairs or cilia on the outside that move it through the water. During the later stage of infancy it turns outside in and becomes attached to something solid. The hairs once on the outside are now within, and so waft life-giving water into it. The water enters through small openings and flows out large ones. Most sponges are colonial so there are a great number of intakes, but the outlets are large so that the outpouring will not be held back. A sponge can eat only microscopic things in the sea.

Strange Creatures of the Deep

ANEMONE

The anemones when disturbed withdrew into themselves squeezing out most of their moisture in a fraction of a minute, and instead of a lovely flower-like creature becomes a warty or soft slimy blob.

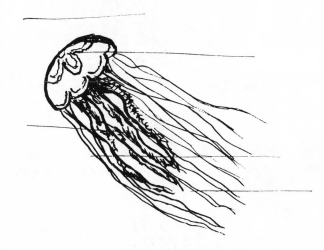

JELLY FISH

The next creatures in the ladder of progression are those that bear the stinging cells — the jellyfish, sea anemone, hydroids, and corals. Jellyfish and only the group to which they belong have stinging cells for their protection as well as for securing food. Most of this group have two forms, free swimming that lay eggs and sessile forms that develop from the eggs to grow into plant-like forms that bring forth the free swimming forms. This turn and turn about of forms is called alternation of generations. Of the jellyfish only a few are sessile when mature but those are capable of crawling.

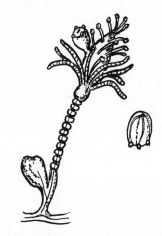

ALTERNATION & GENERATIONS

Hydroids have alternation of generations as shown — both a completely sessile plant-like form and a free-swimming bell.

Joint-legged animals — the arthropods — are the most varied and numerous of all creatures. They abound in the water, the land, and the air. We know them as fossils from the ancient Cambrian Seas recorded there, hundreds of millions of years ago where sedementary rock layers have preserved the fact of their being by imprints and trails as clear to the student in those great layers of rock as a pressed flower between the pages of a book. In all the groups or phylums there are always one or two members that do not conform to the general pattern.

AMPHIPOD

In the great family of arthropods — the joint legged — there are odd narrow little leapers and side swimmers, the amphipods, and flat creepers like pill bugs, the isopods. These are all small.

ISOPOD

Worms and Barnacles Most Common

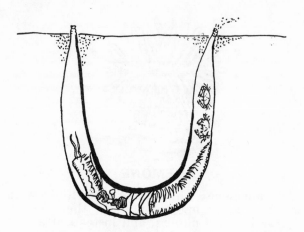

PARCHMENT WORM

The worms of the sea are so different from those on the land. There are the flat worms, like tiny animated oval rugs, round worms like threads and strings, and those that are segmented like our earthworm. Segmented worms are the most commonly picked up and many are very beautiful.

The segmented worms live at all depths. The parchment worm, living in a buried leathery U-shaped tube, obtains its water through the protruding chimneys. It must live in soft sandy mud between tides, or just below low tide. The average tube would be about 14 inches long.

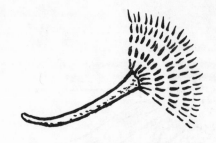

DUSTER

The delicate little feather worm, two inches or less in length, creates a slime tube at the bases of weed and crevices in grassy areas.

CARNATION WORM

But the beautiful carnation worm lives in a crested twisting calcareous tube from wharf piling to great depths.

ACORN BARNACLE

Barnacles are first in this group, and though most wear shelly coverings there is a naked barnacle, and barnacles that are sessile once they begin to mature. As you will see in the text on barnacles some sit flat, others hang suspended.

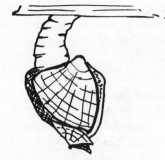

STALKED BARNACLE

TRUE SHRIMP FORM

All shrimp are not at all like the edible shrimp, though they all follow the same general pattern.

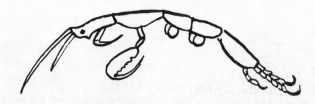

SKELETON SHRIMP — One of the exotic types.

There are many side issues like the skeleton shrimp.

Crabs Take Wierd Forms

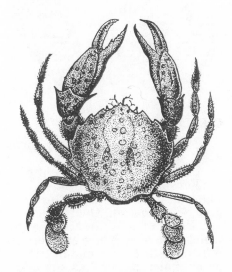

Then there is the true crab form — some equipped with walking legs, and some with the rear legs fashioned for paddling.

As strange a group as any are the starfish family. The five members of this phylum have common characteristics found in all, but the variation in body pattern is amazing. They are all of a radial pattern, easily seen in the starfish, the brittle star, and the sand dollar — and urchin, but the sea cucumber is not covered with calcareous plates as they are, but with a tough leathery skin in which separate plates lie buried.

STARFISH

SAND DOLLAR

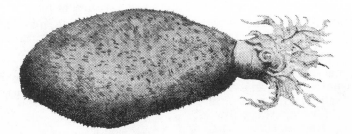

SEA CUCUMBER

The fifth form, the sea lily, wo do not see inshore. The book closes with the lower chordates, creatures that have no vertebral colums, but have a dorsal nerve cord. From here on up animals developed vertebrae.

Minute Animals

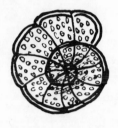

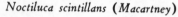
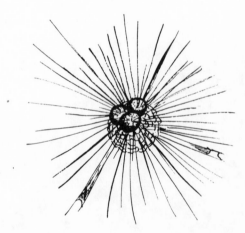

CHALK ANIMALS
Foraminifera

Minute animals, barely visible to the naked eye.

Crystalline or chalky white.

Foraminifera are best seen under a hand lens. They may be crystalline or chalky — simple or many-chambered. Over the surface of the shell are tiny holes — hence the name foraminifera which means hole bearer. The animal fills the shell putting out long slender filaments through the many holes as well as the opening.

Though these creatures are so tiny as to be barely visible, the great chalk Cliffs of Dover, in England, are made of their dead shells.

All chalk beds once lay beneath the sea.

Found throughout the oceans of the world. A handful of mud or muddy sand along the shore will contain many.

NIGHT LIGHT
Noctiluca scintillans (*Macartney*)

Size 1/16 of an inch.

Almost transparent.

These tiny creatures cause the phosphorescent seas. At night the noctiluca rise from the deeper water where they have spent the sunlit hours. Any disturbance in the water causes the noctiluca to glow with an eerie blue light—every fish and wave is a streak of fire.

Each little animal is round with a short flagella for securing the diatoms on which it feeds. It floats like a miniature balloon, incapable of active movement, but by a strange process can rise or sink in the water.

Found in all warm seas.

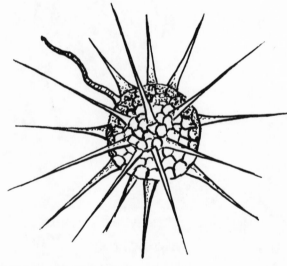

JEWELS OF THE SEA
Radiolarians

Minutens scarcely visible to the naked eye.

Yellowish brown, but pale.

Radiolarians contain small plant cells that give them their color. Because of these plant cells they must come up to the sunlit areas. In exchange for the carbon dioxide given off by the animal, the plants return oxygen. Beneath the elaborate skeleton is a frothy interior. Radiolarian skeletons are made of silicon which they extract from sea water as the globigerina extract lime.

Radiolarian skeletons are the polishing agent in Tripoli stone.

Found in shallow and deep waters.

Globigerina dubia and G. rubra

Barely visible to the naked eye. 0.75 m.m.

G. dubia is white or colorless.

G. rubra is pink.

Globigerina have chambered shells. G. dubia has 50 or more chambers. Lives in deeper offshore waters, but found in shore mud. G. rubra has 3 chambers. Lives from deep offshore waters to shallow waters just below low tide. The shells of these tiny animals are made of lime extracted from the sea.

Most of the foraminifera in the world today are globigerina. As they die their shells rain down steadily to the ocean floor. Three-tenths of the gray mud of the ocean floor is globigerina ooze. Through untold centuries these skeltons have remained beautiful and individual.

Found in all oceans of the world.

Porifera
SPONGES

Sponges are animals, so plant-like in their growth, and so inert that for years they were thought to be plants. Then it was discovered that sponges laid eggs.

Each begins life a small individual. Some small sponges are always individuals, but most of them are great masses of minute sponges so tightly packed together that one cannot be singled from the whole.

The newly hatched sponge swims with minute hairs on the outside. It everts, sits down on a suitable place and becomes attached. The hairs, called cilia, are now on the inside. Constant movement of the cilia keep a constant stream of water passing through the sponge that it may eat and breathe.

Sponges may follow one pattern of growth in an environment, and a completely different one in another. The only way to determine a sponge's true identity is by the inner framework or skeleton. These skeletons are either silicate or calcareous spicules, or horny spongin. Only those with horny skeletons are used commercially.

The only known material that can be used in the glass works for wiping hot glass is sponge, for it will not burn.

Within the passages of sponges live many shallow water creatures too delicate to withstand the wash of the waves.

Sponges are studied and classified, but little is known of their life history.

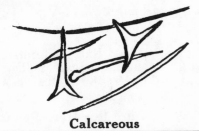
Calcareous

Silicious

Horny

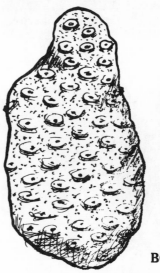

TUFTED SPONGE   *Grantia ciliata* (*Fabricus*)

Also called urn sponge.

Average — ¾ to 1 inch high. ¼ inch in diameter.

Small groups of tiny finger-like sponges — cylindrical and elongate.

One of the solitary sponges, hairy in appearance with slender shining spicules like bristles ringing the osculum at the summit.

Found in pools and shallow sheltered waters on shells, stones, and seaweeds from the Arctic south.

Maine to Cape Cod and south, and North Pacific.

BORING SPONGE
Cliona celata (*Grant*)

Also known as sulphur sponge.

Average boring larva is ⅛ inch across. Sponges are 10 to 12 inches inshore, and great boulder-like sponges in deep waters.

Various shades of sulphur yellow, darkening with age.

Sulphur sponges begin life as larva, bore ⅛ inch holes in shells, penetrating them in all directions consuming the substance bored into. Protruding through the surface, they grow in thick, irregular masses 8 inches or more in diameter. Their real business is to help rid the ocean bottom of waste shell, however, these sponges do bore into living oyster shells, and destroy limestone structures below low tide.

Common in shallow to moderately deep waters in all warm seas.

9

Porifera
SPONGES

FIG SPONGE   *Suberites ficus* (*Johnston*)

Height 3 inches.

Ruddy or reddish brown.

Bulbed like a fig. Skin fig colored, fine and smooth with few oscula.

Usually attached to dead shells by narrow stalks.

Sable Island to Britain.

GRAY SPONGE   *Suberites undulatus* (*George & Wilson*)

Size of hand, 2 or 3 inches high.

Light gray.

Formless, unbranching, small flattened lobes crowded at top. Often attached to edges of dead shells and live oysters. These sponges are very fragile.

Comon in muddy pools at low tide from Maine to Carolinas.

FIRE SPONGE   *Tedania fibula* (*Gray*)

Three to 5 inches high.

Bright fiery red.

Either chimney or incrusting. True form incrusting. Badly crowded oscula form chimneys. Not only is this sponge fire colored, but stings like fire. The spicules break off and become embedded in whatever they touch.

Warm seas.

Somewhat flattened fingers rise from a short stout stem. The surface is velvety with conspicuous 1/10 inch oscula.

Offshore from Arctic to New England.

DEAD MAN'S FINGERS   *Chalina arbuscula* (*Verrill*)

Average height 8 inches.

Largest 18 inches, perhaps larger.

Grayish white when alive—yellow when dead.

Profusely forked branches from a main stalk. Branches slender and fine textured. Distinguished from C. oculata by finer texture.

Shallow water in all ranges.

10

Porifera
SPONGES

BASKET SPONGE *Hircinia or Spongia canaliculata* (*Laubenfels*)

Also called vase or cup sponge.

Average size inshore, 5 to 10 inches. In deeper waters this sponge grows immense; the largest of basket sponges. When living, the animal is black. When cleaned it is yellow. Being a horny sponge, it is used commercially. The inside is smooth; the outside is covered with rough ridges.

Found in shallow Atlantic waters offshore.

EYED FINGER SPONGE *Chalina oculata* (*Pallas*)

Also called finger sponge.

Average height 10 inches. Can attain 25 inches.

Upright bushy colonies. Beached specimens are dull yellow to dead white.

Iceland to Labrador.

AMERICAN GLOVE SPONGE — EUROPEAN BATH SPONGE *Euspongia officinalis* (*L*)

Average 5 to 6 inches.

Deep brown with horny skeleton.

Great variation in shape—rounded, tubular, or cup. Texture soft and silky. Used in pottery business. Least valuable of commercial sponges. Surface usually covered with soft tufts.

Other varieties of Euspongia are the horse sponge — large, coarse and ridgy; sheep-wool, a soft silky yellow sponge used as a bath sponge — the most valuable; grass sponge — one of the least valuable.

Found in shallow waters in warm seas.

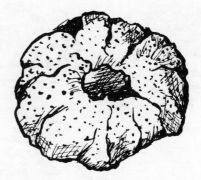

THE WANDERING SPONGE *Tetilla mutabilis* (*George & Wilson*)

Called mud sponge on east coast.

Large as a fist.

Dirty yellow, purple, or red with green highlights.

This sponge is the "tumble weed" of the sea. It rolls about. At times it is loosely attached to eel grass, or it lies idly on mud flats — strange behavior for a sponge.

East and west coasts.

Porifera
SPONGES

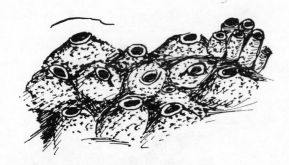

CRUMB-OF-BREAD SPONGE   *Halichondria panicea* (*Johnston*)

Usually a low incrusting sponge 3 to 4 inches in diameter. Changes shape with environment.

Pale greenish yellow, gray, or orange.

Forms spongy almost transparent carpet with frequent oscular openings. Where exposed to the sea, it is flat and paper thin, the more exposed the flatter; where sheltered, it forms thick masses with cone-and-crater surface; from overhanging rocks, it puts out long fingers.

In overhanging rock pools from the Arctic to Rhode Island.

CUP SPONGE   *Phakellia ventilabrum* (*Johnston*)

Average size 6 inches wide and 6 inches high.

Pale yellow tinged with green.

A thin cup-like sponge supported on a slender stem. The young often fan-shaped or branched. Codium, a sponge-like green seaweed may attach itself within the cup.

From Brazil to Great Britain.

RED BEARD   *Microciona prolifera* (*Verrill*)

Height 6 inches. A silicious sponge.

Young bright red darkening to deep orange-red.

Red beard is the crimson coloring on walls and overhangs of rock pools in shallow water. Not found below 60 feet. There is much variation in shape. Common at low water mark.

This is the sponge that is being used to study regeneration of parts.

All moderate to warm waters.

STALK SPONGE   *Stylocordyla borealis* (*Loren*)

Colored orangy.

About 2 inches in diameter.

An odd sponge with a globular head topped by a single osculum and supported on a slender stalk. Long spicules radiate from the center.

Common in offshore waters of North Atlantic — stragglers in south. Seldom washed ashore — brought in by fishermen's nets.

Porifera

SPONGES

RED ENCRUSTING SPONGE *Isociona lithophoenix*
(*General Group Hallman*)

Layers ¾ inch thick.

Fiery red.

Soft lumpy sponges with small oscula set in depressions. Cover the walls of granite caves and rocks.

Vancouver to Laguna Beach.

TUBED SPONGE *Reniera tubifera* (*George & Wilson*)

Average mass about 6 inches in breadth. The tubes are about ⅓ inch in diameter.

Color in life pink or reddish brown to brown. Dead sponge is pale dirty gray. Grows in tangled masses.

Found in shallow, protected waters. Hard to the touch, but very fragile. Often washed ashore in warm coastal waters.

East coast.

URN SPONGE *Reniera urceolus* (*Rathke & Vahl*)

Average height 5 inches or less.

Drab.

A strange urn shaped sponge that has a thin skin with a soft wooly appearance. At the summit is a very large osculum that runs all the way through the body.

Native to Iceland and all North American coasts.

ORANGE SPONGE *Tethya hispida* (*Oken*)

Average 1½ inches in diameter.

Color orange.

Form and color of an orange. Even the thick leathery skin resembles that of an orange. Bunches of glassy spicules radiate from the center. Often washed in around the Gulf of Maine and occasionally as far South as the Gulf of Mexico.

At times small warts appear on its surface. These grow round and rise on stems, finally opening into star-shapes. These are buds — non-sexual reproductive bodies. Offshore waters.

Porifera
SPONGES

Leucoselenia botryoides (*Fleming*)

Height 1⅓ inches.

Yellow or white.

A family of small colonial sponges growing massed or singly. Always in shallow waters below low tide, on solid supports such as stones and pilings.

North east coast.

LOGGERHEAD SPONGE   *Speciospongia vesparia*

Barrel-like sponges up to the size of a tub.

Black in the center fading out at sides.

Large barrel-like, bulky sponges growing close to low tide. Thousands of small creatures live within the oscula. In return for sanctuary, they keep it clean. These small creatures cannot withstand the wash of the shore, yet must live in shallow water. Inert though the sponge may be, it responds to food by a wild lashing of the cilia within.

WESTERN SULPHUR SPONGE
Verongia thiona (*Lambe*)

Three to 4 inches high.

Yellow in life. Purple or black when exposed to the air.

Very coarse texture. Oscula few.

West coast.

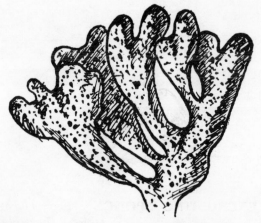

PORITES   *Porites porites* (*Pallis*)

Porites are great masses of reef building coral.

Polyps golden yellow.

At first porites is incrusting, then great knobby, loppy branches form, always reaching up toward the light. The mass is said to increase 32" in 23 years.

Found from Florida to the West Indies.

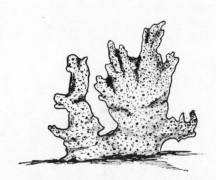

PEPPER CORAL   *Millipora alcicornis* (*L.*)

Also called fire coral and stinging coral.

This is not a true coral.

Grows in great masses helping in the formation of reefs.

Color light brown.

At first millipores form thin encrustations, then spread out with no particular design of growth. As they grow up they reach out coarse stubby-tipped fingers. They are related to the little hydroids, but have calcareous skeletons. The skeleton is 99% of the mass.

They are found in coral reefs over the world in association with porites. There are two distinct types of polyps on pepper coral. A dried finger of coral shows larger openings surrounded by 5 to 7 smaller ones. The feeding polyp is in the center. Those in the ring around it are stinging, food-getting polyps. The small ones have no mouths, but have batteries of stinging cells so powerful that they can penetrate human skin and inflict dangerous "burns".

Coelenterates

HYDROIDS

Hydroids, gorgonians, jellyfish, and sea-anemonies are jelly-animals—flower-like and delicate, but stinging. All have firm jelly-like substance that is 98% water — all live in warm seas. They may be small like the corals and hydroids, or large like certain enemones and the giant jellyfish.

These are the lowest form of animals with body organs — digestive, reproductive. There is sufficient muscular and nervous system for coordinate movement.

Coelenterates have two types of body form — sessile or attached like plants, and medusoid or free swimming like jellyfish. This is known as alternation of generations. Sessile forms rest mouth up, medusoid swim mouth down, but which ever form they wear, jelly animals are always predacious. Sessile forms eat and grow to produce the medusae that lay eggs that grow into sessile forms.

The bodies of coelenterates are sac-like—open at one end and closed at the other with a single internal cavity that is all-useful. There is but one body opening — a mouth. At the closed end is an attaching mechanism.

Around the mouths of coelenterates are tentacles equipped with stinging cells called nematocists, for their protection and for securing food. The tentacles alone could not capture food. No other animals have these stinging cells. A nematocist is like a small filled flask with long slender tube for a neck. Sometimes the tube is tipped with a small spine. When not active, the tube is coiled, outside in, within the cell. When touched a delicate trigger operates to shoot the tube, outside out, to penetrate whatever it touches and inject a paralyzing drug. If the nematocists inject a sizable prey, the tentacles carry the stunned animals to the coelenterate's mouth, but should a predator come within reach of the tentacles such an unwelcome assault would likely drive it away.

A single tentacle may have thousands of these stinging cells. Some are harmful to man, so it is wise to avoid them if larger, or handle the smaller ones carefully.

Coelenterates have innumerable offspring, but only those that settle in a favorable place will ever survive.

HYDROIDS

Hydroids are small — all are best studied under magnification. They are polyps like anemones, enclosed in a skeletal covering.

Hydroids are coelenterates that multiply by budding and by medusa. Those that bud grow into fern and bush-like forms, or into flat encrustations like mosses on rocks and shells — a few are lone like delicate flowers.

When these animals bud, the buds do not break away, but remain fixed to form colonies. In each colony are individuals that feed, others that protect, and those that reproduce sexually.

All the hydroids in a colony are connected by a system of tubes so that the food obtained by the feeding polyps may reach all individuals — thus, if only a few eat, none go hungry.

Hydroids are found hanging suspended from ledges and boulders — on pilings, gulfweed and other seaweeds. One kind lives on the red abalone.

These creatures prey upon the living and in return are preyed upon. Certain sea slugs eat them bodily. Some sea spiders feed upon their juices. So great are their numbers that no figures could show them.

PLUME HYDROID

Plumularia setacea (*Ellis*)

Average ½ inch long.

Largest 2 inches.

Glassy. Hard to see.

Most delicate and beautiful of intertidal creatures. Plume-like in appearance. Give rise to planula larvae instead of medusae.

Key West, Florida.

OSTRICH PLUME HYDROID

Aglaophenia struthionides (*Murray*)

Average 6 to 8 inches.

Almost black.

Live in large vertical rock crevices where they are pounded by the surf.

West coast.

Coelenterates
HYDROIDS

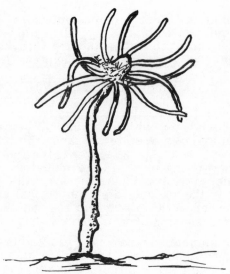

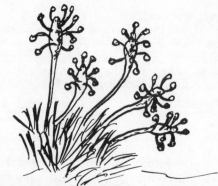

MUD-FLAT HYDROID *Carymorpha palma* (*Torrey*)

Average 3 inches.

Color — pink.

Grow in minute forests. Solitary. Attached to mud-flat with "roots" but unlike rooted plants they collapse when tide ebbs from them, withstand the exposure, and rise with the incoming tide.

Each animal has a flower-like head. On this, small jellyfish form, but never separate from the parent. The jellyfish lay eggs and die. From these eggs the sitting forms rise.

Eaten by sea spiders.

Found in San Diego area, California.

WESTERN PINK CARPET *Hydractinia milleri* (*Torrey*)

Average 1/5 inch high.   Pink.

Spines are smooth. Medusa buds begin to form, but never detach. They mature on the parent and shed sperm and eggs into the water.

Found on sides of rocks.

California to Sitka, Alaska.

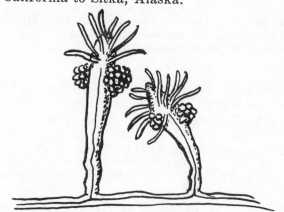

CLUB HYDROID *Clava leptostyla* (*Agassiz*)

Average ¼ inch.

Male — pink to red. Female — purple.

Spreads by creeping under sand with grass-like stolons. No free swimming form. Larvae settle down close by.

Live in cold water.

Long Island Sound to Labrador.

PITH HEARTED HYDROID *Tubularia crocea* (*Ag.*)

Average 1 to 2 inches high.

Delicate pink and red heads.

Long stemmed — in masses. Live on pilings, floating docks, and submerged logs.

Grape-like clusters are reproductive organs. They give forth active larvae that settle down to form new colonies.

Eaten by sea spiders.

Bay of Fundy to Palm Beach, Florida and west coast.

WREATHED HYDROID *Sertularia pumila.* (*Linn*)

Average 2 inches. Buff colored.

Branched forms rise from both sides of a stiff creeping stem.

Grow in tufts on seaweeds. No free swimming medusa.   New England to Carolina.

Coelenterates
HYDROIDS

PINK CARPET *Hydractinia echinia* (*Fleming*)

Also called spiny hydroid.

Average height 4/10 inches.

Males pink, females red or orange-red when expanded.

These hydroids spread extensive roseate carpets on submerged peat beds· They form a fuzzy coating on the shells of the hermit crab, Pagurus longicarpus.

Beneath the colony is an intricate web of stolons. In additon to nutritive, defensive, and generative polyps, pink carpet has some that are sensory.

Each colony contains but one sex.

There are no free swimming medusae.

Found from Labrador south.

SILVERY SERTULARIAN *Thuiaria argentea* (*Linn*)

Average 1 foot.
Largest 18 inches.
Color—silvery on dark stem.
Attach on upright surfaces and float gracefully out in the water.
New England coast.

PINK HYDROID *Pennaria tiarella* (*Ayres*)

Average — 4 to 6 inches.

White to bright pink. Stems yellow to black.

Colony branches are feathery· Medusae bud out of head — ⅛ inch across. They are shed at night, swim around the pilings and settle down near the parent.

Found on eel grass, stones, gorgonians, wharves, and wood below low water mark.

In clean strong flow of water in all ranges along east coast.

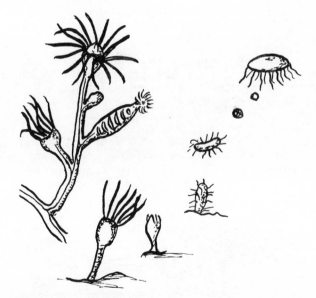

HARBOR HYDROID *Obelia commissuralis* (*McCrady*)

Also called bay hydroid.

Average size 8 inches high.

Grow in bushy colonies with both polyp and medusa. Polyps give rise to medusae which lay eggs and die. Eggs hatch into larvae which swim awhile, then settle down on rocks or seaweeds, attach one end and grow into polyps at the other —so the colony begins again.

New England to Florida.

17

Coelenterates
JELLYFISH

Beautiful jellyfish, swimming by rythmic opening and shutting their bells, jet themselves slowly through the water — so slowly that they are at the mercy of the winds, waves and currents. Some rise to the sunshine and sink with the shadows; others swim by night, some in the dark, some by moonlight.

These lovely translucent jelly-like creatures are mostly water, turned to strong jelly by admixing with sea-salts and organic substances. Though they have no supporting skeletons, they do have nerve fibers and muscles around the edge of the bell for contracting and expanding.

The body of the jellyfish is shaped like a bell, an umbrella, or a mushroom. Beneath the bell, in most jellyfish, hang long tentacles studded with powerful stinging cells. Those few that lack tentacles have stinging cells scattered over the bell. These stinging cells are for capturing prey, but can do damage to man. Some have such power as to be very dangerous. Jellyfish have eye-like organs along the edge of the bell, but Nature is stingy with eyes in the ocean and as most of the jellyfish have little control over their direction of travel, it is doubtful that they see more than light and darkness, or shadows.

The cross or horseshoe markings at the top of the bell are the gonads or sex organs.

Jellyfish are short lived, carniverous animals seldom living more than a year. Very few live inshore. Many that are washed ashore are pitifully unlike the lovely jellies of the open sea.

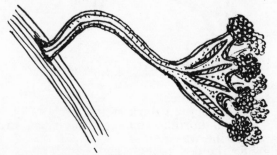

STALKED JELLYFISH   *Lucernaria quadricornis*
(O. F. Miller)

Also known as many-mouthed jellyfish.
Two to 2½ inches high.
Gray-green to brown hues.
These jellyfish are sedentary, but not permanently fixed. They are quadrangular in shape with 8 tentacles. Like hydroids, they sit with bell upside down — mouth up. The tentacles are hollow knobs in groups like pompoms. They are considered an undeveloped form. In the notches between the tentacles are adhesive pads that can be used as anchors if the jellyfish is knocked from its footing.
Found on seaweeds north of Cape Cod.

MANY MOUTHED JELLYFISH   *Haliclyatus salpinx*
(Clark)

One inch wide and ¾ of an inch high.
Color — subtle tints of a soap bubble.
An upside down bell, anchored by a slender stalk. Though it has 8 groups of tentacles it is four sided in cross-section.
Found on the rocky New England coasts.

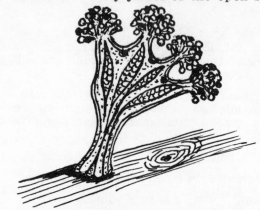

ATTACHED JELLYFISH   *Haliclystus stejnegeri*
(Kishinouye)

One inch in diameter.
This is related to the large jellyfish such as are seen well off shore. It has lost the power to swim, but can contract slightly and fold in its tentacles. If loosened it will re-attach. It can shuffle along like a snail.
Found on eel grass from Puget Sound to Alaska.

ROOT-MOUTHED JELLY   *Stomolophus meleagrs* (L.)
Also known as many-mouthed sea-jelly.
Usually no more than 7" across.
The bell is a milky blue or yellowish with brown reticulations all over, becoming very dense at the margin, and so giving it a deep brown tone. It is thick and rigid at the margin. Shaped somewhat like a peaked mushroom with no tentacles. The thick-walled mouth arms form a stem-like mouth. It is a strong swift swimmer.
Found from North Carolina around Florida to the Gulf, and in the Pacific.

Coelenterates
JELLYFISH

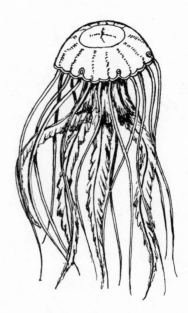

LAPPET BORDERED JELLYFISH
Periphylla hyacinthina (*Steen*)

Six inches high, 5 inches wide.

Reddish purple inside bell. Lappets pale translucent blue.

A funny high-pointed bell with gracefully curving sides. There are 16 marginal lappets and pouches. A deep furrow lies between the bell and the lappets. There are twelve tentacles and four sense organs.

Found in the North Atlantic up as far is Greenland.

SPECKLED JELLYFISH
Dactylometra quinquecirrha (*Ag.*)

Also known as the golden furbelowed jellyfish.

May be eight inches to one foot across. Forty golden yellow tentacles hang down below the rosy lappets.

The sting is most painful.

In temperate to tropical waters in all seas.

LAPPET BORDERED JELLYFISH
Nausithoe puncata (*Koll*)

One and one half inches across. Very flat.

Translucent green or light brown with reddish tones on lappets. The gonads are red, brown, or yellow. A furrow separates the lappets from the bell. The tentacles bristle out almost straight from the bell.

These jellyfish float at the surface in all warm seas.

LUMINOUS FURBELOWED JELLYFISH
Pelagia cyanella (*Peron & Les.*)

Two inches across.

Rose-purple bell shading into blue, dotted orange-red nematocysts.

It is a very active swimmer with a strange swimming movement that makes it first a ball and then a saucer. The nematosysts on the disc are in long radiating red lines. The sense organs are between the tentacles. This jellyfish can give a painful sting. It is luminous at night.

They wash ashore in late summer from Cape Cod to Florida.

Coelenterates
JELLYFISH

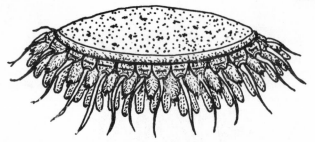

ATOLLA
Atolla bairdii (*Fewkes*)

One to 6 inches wide with 22 tentacles.

Translucent milky blue flecked with red. A red ring under tentacles.

A short four-lipped mouth in the center.

Found from the Gulf to North Carolina, surface to depths.

SUN JELLY
Cyanea capillata, arctica (*Esch.*)

Also known as the great pink, lion's mane, red jelly or sea blubber.

Often grow three to four feet in diameter, but far out at sea they are eight feet across. They are largest in cold water.

The stomach is rosy pink — the gonads are reddish brown. The edge of the yellow or blue bell has 6 to 800 seventy-five foot tentacles hanging from it.

When the tentacles are spread they are 25 times the diameter of the umbrella. Four long thin folds hang in the center with such full ruffly edges that they seem almost like a petticoat.

These great creatures are animals of warm summer seas that adult in a single season.

The sting is horrible, yet within the protection of its tentacles, the young of our butterfish find sanctuary.

A bluish pink variety on west coast.

C. versicolor of the west coast is a bit smaller and usually yellow or orange. It eats moon jellies, crabs or anything floating.

Found off shore from Greenland to North Carolina.

MOON JELLY
Aurelia aurita (*Lam.*)

Also called the white sea-jelly.

Average size 1 to 10 inches. A 24 inch one is very large.

Moon jellies are colorless. The male gonads are pink and violet; the female brown or yellow opalescent.

Within, the central column is a four-sided mouth with which it captures diatoms and small larvae. Though they sting, it is only to capture food — they are harmless to man. Gravity sensitive organs on the rim cause it to rise whenever it sinks.

The colder the water, the thicker the jelly. In the fall, towards the end of life, they often occur in great swarms just off shore. Often great masses of wave-torn jellies are washed up on the beaches.

Moon jellies are found in all coastal waters of the Atlantic and Pacific, from the poles to the equator.

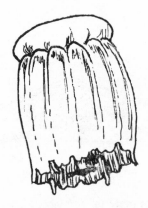

THIMBLE JELLYFISH
Linuche unguiculata (*Escns.*)

One half inch high 3/5 of an inch in diameter.

Pale blue.

The constriction is near the top of the small cylindrical umbrella. There are 16 grooves on the sides with 16 blunt lappets and 8 very small tentacles.

Found in the Gulf Stream and often very numerous near shore.

Coelenterates
JELLYFISH

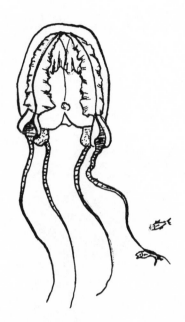

FOUR SIDED JELLYFISH

Tamoya haplonema
(F. Muller)

About 3½ inches high 3 inches in diameter.
Bell transparent. Tentacles yellowish with tinges of purple. Organs yellow.

There are four tentacles with leaf-like spines. Between the tentacles the bell is flattened, giving the animal a four sided appearance. In the notches between the tentacles are one large eye and two small ones. The stomach fills almost the entire bell.

The stinging cells or nematocysts are in wart-like clumps along the margin of the bell.

Found in all warm seas from New York to Brazil and the West Indies.

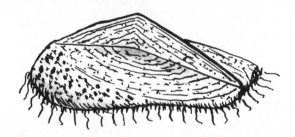

BY-THE-WIND SAILOR

Velella mutica (Bosc.)

Velella mutica of the east coast and V. lata of the west coast are known as "by-the-wind sailors".

Velella has a small crisp sail that is set diagonally across the exquisitely delicate blue-green float. Short blue tentacles fringe the body which is seldom more than four inches long.

Both forms live in warm seas.

V. mutica seems to be free of uninvited guests, but V. lata often trails purple zooids beneath it, and sometimes a small milky-blue goose barnacle travels on it.

PORTUGUESE MAN-OF-WAR

Physalia pelagia (Bosc.)

A very few creatures have attained the power to become lighter than water. The siphonophores are among them.

The best known of these is Physalia pelagia the Portuguese Man-of-War — a stinging jellyfish that wears a beautiful blue float, topped by a rose crest. Beneath the float hangs a colony of animals. The groups of short salmon-pink ones are for reproduction, the delicate blue trumpet-shaped organs are for eating, the pale green palps are sense organs, the long lavender-blue ribbon-like tentacles are for protection and capturing prey.

Physalia travels with tentacles spread wide to capture any unfortunate that passes. These tentacles possess the most powerful stinging cells of all marine animals, and can be dangerous to man.

In its voyaging Physalia is subject to winds and strong currents. but it has some control over its movements, for the float is filled with gas supplied by a specialized gland, and it continually changes shape as it trims itself to the wind. It can sink at will by expelling gas, and rise by regenerating gas.

A fish captured by the tentacles is paralyzed and drawn up to the many mouths which suck it dry. There is one small fish, however, that swims unharmed within those dread tentacles. The small fish lure larger ones within reach of the tentacles and, as Physalia eats, they gather up the crumbs from under the table.

Coelenterates
SEA ANEMONES

Sea anemones are beautiful, fragile, and flower-like, but their beauty belies their true nature. They attach themselves to something solid above or below the sand, patiently awaiting any hapless creature that blunders against their tentacles. Nematocists fling their paralyzing arrowheads into the victim — the exquisite tentacles curl over it bringing it into the waiting maw. Again the delicate tentacles open and the small tragedy is overspread with beauty. Voracious it is true, but all must eat. The anemone will try almost anything, but if it is not suitable food will reject it.

Anemones grow in proportion with their food. They even shrink with lack of it.

In tidal areas when they are bared for so brief a time they withdraw into themselves by elimination of water from their tissues and become wrinkled unattractive masses.

Though usually attached, anemones can creep about like snails, but at the snailest of snail paces.

Within the shelter of the tentacles of some species, small fish seek protection, and in return lure unwary victims within reach of the anemones.

GREAT BURROWING ANEMONE
Cerianthus americanus (*Verrill*)

Length 6 inches.

Dark brown — 125 tentacles in each of two circles. The inner row short, the outer row long.

Lives in shallow water in a black parchment tube.

Found from Cape Cod to Florida.

The great burrowing anemone on the west coast is Cerianthus aestuari. It also may be 6 inches long. It is mottled brown with banded tentacles. There are 60 tentacles, the outer ones are long — the 30 inner ones are short. They, too, live in a black parchment tube covered with muck, but the inside is satiny. Found mostly in shallow water.

SLIME TUBE ANEMONE   *Cerianthus borealis* (*Ver*)

Length 7 to 9 inches. Width 2 inches. Expanded tentacles 5 to 6 inches across. 150 to 200 tentacles.

Color variable. Chocolate brown to orange brown, green or gray. The disc is pale yellow.

This is the flower-like anemone of the east coast. It has a long, rough, thick tube made of sand grains glued together to make a slime tube. The lower end has a round terminal pore. It lies with its crown of tentacles exposed on the sandy mud.

It lives on sandy shoals, estuaries, and mangrove inlets.

Found from the Bay of Fundy to the Gulf of Mexico.

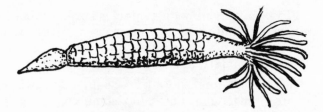

EDWARDS ANEMONE
Edwarsiella sipunculoides (*Stimpson*)

Small - 2 inches long, 20 to 36 tentacles, 4 inches across when extended.

The tentacles are transparent brown.

This anemone has no attachment disc. It has a pointed foot for burrowing. The body is covered with a wrinkled brown tube with 8 longitudinal ridges which give it the appearance of a rather worn corncob. When young it is often parasitic on jellyfish.

Found on the west coast and around Cape Cod on the east coast.

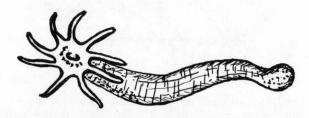

EDWARDS WESTERN ANEMONE
Edwardsia california (*McMurrich*)

Average 2 inches in length.

Milky brown. This is a burrowing anemone with 8 longitudinal bands on the body. When open, the tentacles are a little above the bottom; when dug out they contract violently.

Found in profusion on mud and sand bottom, only in southern California.

Coelenterates
ANEMONES

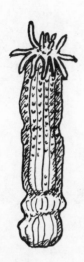

TEALIA   *Tealia crassicornia* (*Muller*)

Height 2 inches, diameter a little over 4 inches.

The body is streaked with crimson and dark red, blotched with lighter tones. Bright lines run from the mouth to the tentacles. The tentacles are ringed with red and white.

This is a squat form, with a soft changeable shape. Over the surface of the body are small suckers for attaching to solid things.

Found on both coasts north of Puget Sound and Cape Cod.

KNOBBED ANEMONE   *Eloactis producta* (*And*)

Nine inches long, ¾ of an inch in diameter.

Whitish yellow or gray with 20 long rows of whitish warts.

There are 20 short, blunt tentacles. They frequent the shallows, lying buried in mud or sand up to the tentacles. They are also found under stones. They have no tube. The body is worm-like and **retractile**.

Found from Long Island southward.

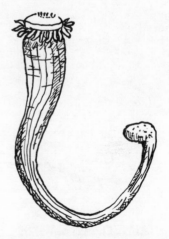

DAHLIA ANEMONE   *Tealia felina or Urticina felina*

Also called beaded anemone, and thick petaled rose anemone.

Diameter 4 inches, of a squat build.

Green and red. There are 160 thick tentacles.

This anemone may either bury itself in the mud or cover itself with bits of rock and shell, or attach to rocks.

They are circumpolar, living at extreme low tide as far south as Maine.

BURROWING ANEMONE

       Harenactis attenuata (*Torrey*)

Ten inches long, 6/8 of an inch in diameter. It can stretch to 18 inches. Has 24 short tentacles.

Sandy gray in color.

This animal builds no slime tube, but burrows in sand and mud, and when disturbed pulls in its tentacles and retracts into the sand. As it retracts the bulb at the lower end swells so large that it cannot be pulled from its burrow.

Found in southern California.

Coelenterates
ANEMONES

ORANGE STRIPED ANEMONE
Sagartia luciae (*Verrill*)

Height ⅓ to ¾ inches.

Twenty-five to 50 long pale green tentacles.

Handsome olive green with 12 longitudinal orange stripes.

They live attached to small stones in tide pools from Massachusetts to Florida on the east coast, reproducing by lateral division or ciliated larvae. Those on the west coast have 84 tentacles of varying length set in four rows about the mouth.

PLUMED ANEMONE
Metridium plumoso (*Oken*)

Six to 8 inches long.

A brownish yellow blob when withdrawn or the tide is out. May be olive-brown, white, orange, peach, or mottled. One thousand tentacles.

The plumed anemone can move slowly about, often leaving crumbs of itself behind which grow into new anemones. They feed on microscopic organisms. They do not like soft bottom and are usually found on wharves, rocks, and in crevices on both coasts.

WHITE PLUMED ANEMONE
Metridium senile (*Calvin*)

Large ones may be 2 inches high and 6 inches across.

May be rich straw-brown or reddish yellow.

In deep water they cluster together on pilings, but small ones are often found on barnacles and on a tunicate. Those in very deep water grow huge.

Found on the west coast from Santa Barbara to Sitka.

BROWN ANEMONE
Metridium dianthus (*Ellis*)

Height 4 inches, width 2½ inches.

A large one may have a thousand tentacles.

The color is variable. They may be white, but the average color is velvety brown. They are always brown in tide pools above low tide.

This is the largest, handsomest, and commonest anemone found on both coasts. The column is velvety smooth, with a wide fluted oral disc filled with slender tentacles. They reproduce by dividing longitudinally down the center, by fragmentation, by eggs, or by budding near the base.

They live from high tide to deep off shore water, on stones, rocks and pilings.

Their distribution is world wide, even to the Arctic Zone.

Coelenterates
ANEMONES

WHITE ANEMONE   *Sagartia modesta* (*Ver*)

Height 2½ inches, width ¾ of an inch. It is large on the east coast and small on the west.

Yellowish or grayish green with dark basal spots. The disc is yellow, the column — pink or flesh color.

The long slender body has an adhesive disc to attach to stones and shells buried in the sand. Only the tentacles are to be seen spread on the sand and grains of sand adhere to the column.

Found in tide pools in shallow water on the Pacific and from Long Island southward.

SMALL GREEN ANEMONE

Epiactis prolifera (*Ver*)

This anemone is small in its southern range, large in its northern. The average small ones are 2/5 of an inch high, and ½ inch wide.

Color usually green.

There are 96 tentacles.

This anemone wears bands of egg pits around the column below the middle. The eggs are retained in the brood pits until they are regular anemones old enough to glide away. They can always move about.

Found on the protected sides of boulders and on sea lettuce, on both coasts.

GREEN ANEMONE   *Bunodactis xanthogrammica*

This anemone goes under many names and varies in size according to locality and depth. It may also be listed as B. elegantissima, or B. Cribina.

They vary in size from 3 inches to 10 inches.

The color is uniformly green because it has a symbiotic algae that lives in it, and for this reason seeks the light. It is a surf-loving animal, mostly solitary, but in some places they group. Big or little, they sting like nettles.

They are found from Unalaska and Sitka to Panama on the west coast.

TRICOLOR ANEMONE   *Adamasia sociabilis* (*Verrill*)

Two and seven tenths of an inch high, 1.6 wide with 500 tentacles.

These are the anemones that attach themselves to hermit crab shells, living commensally with the crabs. This is an advantage to both, for the crab is protected by the stinging cells of the anemone, while the anemone eats at the crab's table. It has an especially expanded base for comfortable attachment. If a hole occurs in the crab's house it is said that the anemone will cover the hole. Those who have studied the crab closely say that when it changes shells it takes its guest along.

Coelenterates

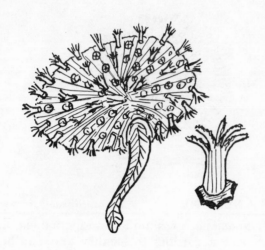

SEA PANSY
Renilla reniformis (*Pallas*)

Fully expanded ones may be 4 inches across, but the average size is 2 inches.

Color may be blue or violet with white polyps on the upper surface.

The sea pansy is a flower-like, heart-shaped animal supported on a short stalk which holds it above the sand, but it can withdraw completely. By night it glows a soft luminous blue. Each pansy, which is a colony, grew from a tiny larva. Like all coelenterates their feeding tentacles reach into the water for living motes — diatoms, fish larvae, eggs, small copepods or any living matter suitable to their taste.

Most of these creatures are infested with a parasitic copepod.

Those on the California coast are Renilla amethystima (Verrill).

RED SEA PEN
Pennatula aculeata (*Danielsen*)

Also called sea feather.

Average 4 inches long.

The head is deep red to reddish purple supported on an orange or yellow stalk. There is a rosy variety, and an albino. The stalk usually pales toward the tip. They are luminescent in the dark.

Found in off shore waters from the Gulf of St. Lawrence to the Carolinas.

SEA PEN
Stylatula elongata (*Gabb*)

Length 10 inches.

Green — luminescent by night.

These sea pens are found in great patches wherever they occur. They are quite securely anchored by a bulbous base on sandy mud flats and when disturbed can completely contract. Most beds of these lie between half and low tide.

West coast between San Diego and San Francisco.

Coelenterates
CORALS

Corals are delicate animals that sit in limy cups of their own making. Some are lone, others grow in small colonies — never more than 20 to 30 individuals to a colony, and some branch and grow into great reefs. Each little coral polyp is much like a tiny sea anemone.

In cold water there are no reefs, for when temperatures drop below 68' corals are depauperate. At 74' they are large and branching, but must have hard bottom. Small colonies and lone corals may settle on shells and small stones.

Corals make their living by reaping the water for microscopic food. During the day they rest, tucked safely down in their stony cups, for they must eat by night as it is then the plankton rises from the deep water.

The closed corals are dull white, but at night, when expanded, they are startling in their beauty — but this beauty is not their own. It is caused by microscopic plants living within the coral tissues, so their beauty is a reflection of the colorful world of plants.

The sexes are separate. Corals bud and produce eggs, which develop in a very short period into gliding, drifting larva.

Our Pacific coast has few corals because of the upwellings of cold water.

The sea's currents shape the coral reef, and at any time it may be reclaimed by the sea.

STAGHORN CORAL   *Madrepora cervicornis*

Huge branching masses that grow 1 to 2 inches a year.

Polyps are yellowish. All color is due to algae living in the corals' tissues.

Staghorn is very porous coral with canals that connect the polyps. Growth takes place at the tips of the branches. A "mother polyp", somewhat larger than the others, sits at the tip and is forever creeping up, building out the tip of the branch as she goes. At her base, polyps bud out and are left behind.   Florida to West Indies.

STAR CORAL   *Astrangia danae* (*Agassiz*)

Also known as "Star of the East".

Polyps 1/4 inch across. Colony, size of a palm.

Polyps white or slightly pink. The cup is brown inside and half way down the outside. They form incrusting colonies of 20 to 40 individuals, usually on stones and shells in shallow water.

Found from Cape Cod to Florida, and from Puget Sound to Monterey Bay.

EYED CORAL   *Oculina diffusa* (**Lamarck**)

These are small tree-like forms, much branched at 30 degree angles.

The polyps are large — generally an eighth of an inch across.

The cups stand out conspicuously from the rather gnarled branches.

Found in cold seas to Florida and the West Indies.

ASTRANGIA   *Astrangia eusmilla* (**Ag**)

Cup 1/2 inch across.

These large lovely corals are brushed with rich brown.

They divide but are attached basaly, always remaining attached.

Found attached to shells, small stones, or on coralline groups, in tropical and subtropical waters.

Coelenterates
CORALS

BRAIN CORAL   *Meandrina sinuosa* (*Lesueur*)

Grow to 10 to 12 feet across, though the average is 4 to 5 feet.

This is a massive hemispheroidal coral. It is solid with only a small outer crust of living coral. Brain corals are reef foundations.

Coral polyps bud and separate from the parent as a rule, some having connecting canals with the parent, but this coral buds and then does not separate completely. So, instead of a polyp with one mouth and a stomach, in the brain coral the long sinuous strings of polyps have but one stomach and many mouths.

Florida to West Indies.

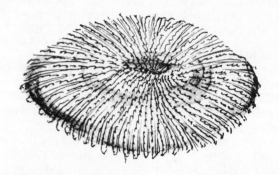

MUSHROOM CORAL   *Fungia elegans* (*Verrill*)

Four to six inches across, on the Gulf of Florida, but on the west coast they are usually ½ inch across.

Polyps cadmium yellow to orange.

Mushroom coral reaches up in regular coral cup formation, then begins to swell at the top, broadening out into what resembles a mushroom turned upside down with the gills showing. This drops off onto the sand and begins to grow. The original polyp gives repeat performances. The shed heads and the mother polyp always remain single individuals.

Gulf of Mexico and Florida and California.

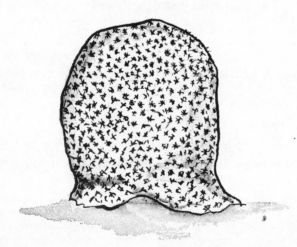

HEAD CORAL   *Orbicella annularis* (*Lamark*)

Large globulose colonies as much as five or six feet across.

The small openings are 1/12 of an inch across.

These are among the great bulky corals that help build the outer coral reefs.

Florida to West Indies.

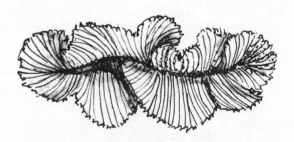

FUNGUS CORAL   *Meanderina meandrites* (*L.*)

Also known as rose coral.

Four to six inches long. Longer than wide.

There is a single main groove down the center from which delicate leafy plates extend to the rim. It is one of the loveliest of the small corals.

Usually found growing at the bases of reefs from Florida to the West Indies.

Ctenophores

Another side-line in the jelly animal world are the Ctenophores or comb-bearers. They mass in incredible numbers and are carried along by the sea. They are here today in swarms, and gone tomorrow. They are inshore surface swimmers, and so, after storms, their tattered bodies may litter the beaches. Ctenophores are blobs of very solid, clear jelly that do not sting — they are not jellyfish, and have no living relatives. The body pattern is not radial like that of the jellyfish, but biradially symetrical.

Ctenophores are quite solid to the touch and incapable of stinging as they have no nematocusts.

Through the body is a single pole with a slit of a mouth at one end and a sense organ at the other. Eight rows of ciliated combs stretch on the outer surface from sense organ to mouth. These combs are like tiny squares attached to a ribbon at one edge and fringed at the opposite edge. Beating a gentle rythm the combs propel the Ctenophore slowly through the water. If there is no sun they are almost invisible, but when the sunlight falls on the combs the ciliated edges break the light into prismic colors, at night the combs are luminous. All Ctenophores are carniverous taking a tremendous toll of larval fish, minnows and small crustaceans.

All are hermaphroditic, shedding their eggs and sperm into the water. There are but 21 species on our coast but their numbers are unbelievably great.

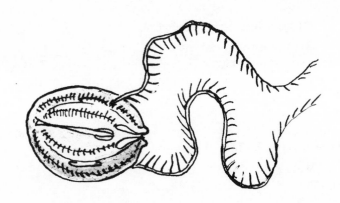

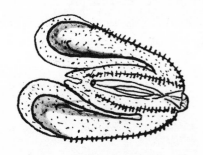

SEA GOOSEBERRY   *Pleurobrachia pileus (Fabricus)*

Also known as cat's-eyes.

Average ¾ inches long, ½ inch wide.

They are feeble swimmers, with a pair of very long delicate feathery tentacles, often twenty times the length of the body and almost entirely retractible in body sheathes. The body is milky white—the tentacles delicate pink to orange-yellow. They swarm along the coast in the summer eating all small animals that come within reach of their sticky tentacles.

They are found from the Arctic to the Antarctic and are most common along our New England shores.

The western form is P. bachei. It is a little smaller than the one on the east coast.

SEA WALNUT   *Mnemiopsis leidyi (A. Agassiz)*

Average 4 inches long.

Color — milky translucent.

Sea walnuts have flap-like lobes for mouths. They swim mouth forward devouring many small fish as well as everything else that comes their way. The oral lobes can draw in giving an appearance of one licking ones lips. Were it not for the ciliated combs flashing in the sunlight it would be almost impossible to see them. However, at night they stand ont clearly due to their high flourescence. M. leidyi is parasitized by a sea anemone, Edwardsia leidyi.

These are the most common of the ctenophores and are found in all ranges.

Sea Worms

FLATWORMS

Flatworms are mostly very flat, wierd little creatures living free swimming, under stones, in old shells, on seaweeds and grasses, and on other creatures. They range from the parasitic to cannibalistic, to vegetarians. Their colors may be beautifully delicate or vivid. Some have mouths where they "belong", while others wear their mouths underneath in the middle. Both male and female are present in an individual, yet they do not fertilize their own eggs.

When disturbed they may crawl away like animate rugs, or take off through the water like miniature flying carpets.

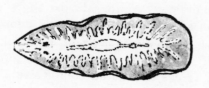

LEPTOPLANA   *Leptoplana acticola* **(Boone)**

Averages an inch in length.

Tan with darker markings.

An active little worm that crawls over stones and weeds in upper tidepools. This is the west coast form.

LIMULUS FLATWORM   *Bdelloura candida* **(Girard)**

May be 3/5 to an inch long.

Gray-white to yellow with brown foliate intestines that shine through the body. It is shaped like a spear-head, broadly rounded behind. There are two closely set eyes. A sucking disc at the posterior end is for attachment. It swims very rapidly.

Found on the gills and outer surface of horseshoe crabs, being commensal, not parasitic.

Maine to North Carolina and rare farther south.

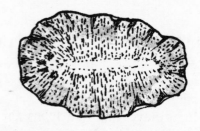

LEPTOPLANA   *Leptoplana ellipsoides* **(Girard)**

One to 1½ inches long.

Yellowish brown with greenish spots and flecks. Light down the center.

The body is oval, flat, and thin. The eyes are in paired clusters. They are wavy swimmers found on the east coast from low water to deep water, usually under stones.

ZEBRA FLATWORM   *Stylochus zebra* **(Verrill)**

One and ½ inches long when full grown.

Body yellowish brown shading to chocolate. Yellow or white tansverse parallel lines give it its name. It is thick down the middle with almost translucent edges. There are close set stubby tentacles with many eye spots which practically meet.

Found mostly in dead welk shells on the eastern coast.

Sea Worms
FLATWORMS

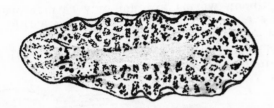

ELIPTICAL FLATWORM *Stylochus ellipticus* (*Girard*)

Not quite an inch in length.

Color—yellow-brown with irregular radiating pattern. There are two white tentacles on the forward end and eight to 10 eyes. They swim by undulating the margins.

Found under stones, in old shells and on pilings.

New England southward.

Polychoerus caudatus

Length 4 mm, width 1.5 mm. Red.

Parallel sides, with a deep indentation at hinder end from which may extend 1 to 3 caudal filaments.

Tidal zones among stones New England southward.

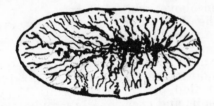

LEAF FLATWORM
Planocera californica (*Heath & McGregor*)

One to 1½ inches long.

Blue-green with black and white markings.

The body is thick, firm and round. They prowl on damp gravel under boulders, eating minute creatures.

Found from Monterey to Lower California.

Phogocata griaeilis

Length 30 mm to 4.5 mm. Black.

Body elongate and flat with rounded ends.

Brackish water.

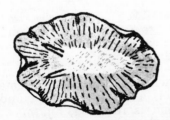

NEBULOUS FLATWORM *Planocera nebulosa* (*Girard*)

One inch long.

Olive green with yellow blotches and a pale dorsal line, two long slender tentacles with clustered eye spots. These worms are thin with delicate edges that undulate as they swim.

Found on the east coast.

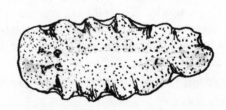

Leploplana variabilis

Eight mm to 18 mm.

Eliptical or leaf form. No true eyes—4 clusters containing from 15 - 30 occelli.

At lower edge of tidal zone, New England southward.

Sea Worms

RIBBON WORMS

Ribbon worms are unsegmented, non-parasitic worms that swim with rapid undulating movements. Their margins are delicate and frilly. There are thousands of species, many of which are brilliantly colored. The ribbon worm has a proboscis, like the turned-in finger of a glove, which can be everted to capture prey. There is often a spine or stinging cells on its tip. This is one marksman that never misses his aim. The sexes are separate. Lost parts are regenerated. They fragment readily.

Ribbon worms are not only carniverous, but many are cannibalistic. However, as a rule, they feed mostly on annelid worms.

These remarkable worms vary in length from one inch to 30 feet.

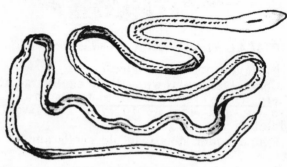

MANY EYED RIBBON WORM

Emplectonema gracile (*Johnstone*)

Average size 8 inches, but can contract to 4 inches.

Green above, white below.

The head is wider than the body. Proboscis is short and thick. Head may have 20 to 30 eyes on each side.

They are to be found between tides lying under stones and tangled among young mussel shells like loosely coiled bits of used string. They inhabit empty shells and occasionally may be found within the shells occupied by large hermit crabs.

Found in shallow water along the west coast.

RIBBON WORM

Micrura verrilli

Length 12 inches.

Brilliantly banded with lavender and white.

These are soft bodied worms that cannot swim. Little is known about them, for they are uncommon.

Bristle worms are often swallowed whole. It is said the bristle worms have mostly acquired the stinging cells in their bristles from coelentetrates they have eaten. The stinging cells are not digestible and make their way into the bristles and live to sting again. The ribbon worms are incapable of digesting the bristles, so the bristles simply find their way out by perforating the intestinal tract and skin of the ribbon worm.

Found on rocky coasts from Monterey Bay to Alaska.

RIBBON WORM

Cerebratulus lacteus (*Leidy*

Largest are 20 feet long and one inch wide.

Flesh to salmon colored down the center bordered with dark brown.

C. lacteus is thick at the front, narrowing and tapering toward the tail with extremely thin edges. The mouth is a long narrow slit underneath. The posterior end is rounded, but may be equipped with a long slender cirrus or sensory appendage. The very thin margins of the worm are well adapted for swimming. There are no eyes. In large ones the slender white proboscis extends three feet.

During the day they are hidden. At night they swim actively about.

Very common in mud and sand at low water mark from Maine to Florida.

WESTERN RIBBON WORM   *Cerbratulus herculeus*

The west coast form seldom exceeds 12 inches. It is dark brown with a white snout. The proboscis is pink. They are found between tides from San Pedro to Alaska.

Sea Worms
ANNELID WORMS

Annulata are segmented or ringed worms with a distinct head. The segments bear appendages, called parapodia, for locomotion. Annelids breathe all over. A proboscis, sometimes coming out through the mouth opening, but mostly through a proboscis pore, works much like a glove finger pulled into the palm outside in. It can be shot out, fully extended, as the blown finger is turned out in a glove. These are handy adjustments for digging and for snatching passing prey. Along the Gulf coast of Florida.

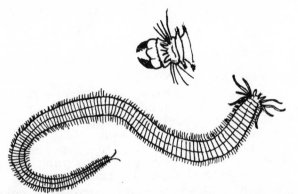

SAND WORM *Nereis virens* (Sars)

Also known as clam worm.

Large ones are 18 inches long with 200 segments.

Body bluish with light green parapodia tinged with orange.

These are the largest and most common worms on our east coast. During the day they hide under stones or burrow into the sand between tides, coming out by night to swim rapidly and sinuously about hunting food. They are extremely voracious. When desirable prey passes, the proboscis is shot out and the two powerful black hooks at the tip make the capture.

These worms lengthen out by adding segments.

Fishermen dig the clam worms out for bait, and certain fish nose them out of the sand. They are cosmopolitan on the east coast.

In the west there are two common forms. Naineris dendritica is found under mussel beds and is commonly used for bait, and Nereis vexillosa usually found in mud. The most remarkable form in the west is Neanthes brandti, often 3 feet long, with many hooks on its proboscis. This one is responsible for some of the sea-serpent stories.

SLIME DUSTER *Myxicola infundibulm* (Montague)

This is a small worm averaging between 1 and 2 inches.

The color ranges from maroon to pink.

The hind end of the thick body is hooked for security. A mucous membrane unites the gills.

Found living in mucous tubes in the mud around mangrove roots, from the Florida Keys along the Gulf coast of Florida.

CARNATION WORM *Hydroides dianthus* (Gunners)

This is a 3 inch worm that lives in a twisted calcareous tube.

The body contains green blood. The crown of gills is long, slender, and radiating. They are purple banded with white and green, or brown banded with yellow and white. The funnel shaped operculum or trap door that closes the opening is topped with several branched setae and rimmed with spines. The upper half of the body is caped.

These worms are common on coral stacks, rocks and shells, usually in fairly deep water from Cape Cod to Florida and the Gulf.

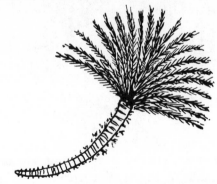

PEACOCK WORMS *Family Sebellidae* (Linn)

Often called feather duster or plumed worms.

These worms vary in size from the dwarf that is but an inch long to gorgeous creatures up to 18 inches in length, and may have as many as 200 segments. The tentacles are for catching warning shadows, for feeding, and breathing by setting up water currents by ciliary hairs.

Peacock worms build parchment tubes—there is usually a part of the tube above the mud.

Various forms are found from the Arctic to the tropics.

Sea Worms
ANNELIDS

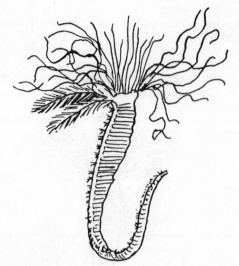

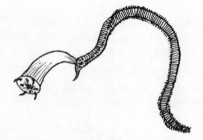

ORNATE WORM   *Amphitrite ornata* (*Verrill*)

Grown worm 12 to 13 inches long.

Flesh colored with brilliant red many branched gills — beautiful and flower-like.

The ornate worm makes its tube in muddy sand areas.

Like many other creatures the breeding is related to the phases of the moon.

They are found at low water mark along the east coast.

BREAK THROWER   *Glycera dibranchia* (*Ehlers*)

Average size 6 to 8 inches.

The worm is pinkish purple with a decided pink line down the middle. The parapodia are pink. The head is small and pointed with no hint of the large red proboscis armed with 4 black hooks. These are for nabbing prey and nipping an enemy. The beak is really the pharanx. There are gills on both sides, under and upper.

The beak thrower is a fine burrower. It pokes its sharp pointed head into the sand and corkscrews down.

Found along the Atlantic coast.

A western form, G. americana, has branched gills on the top only. It inhabits eel grass beds. The fearful proboscis is armed with 4 hooks, and is ⅓ as long as the body. G. robusta, also of the west coast, lives in foul mud.

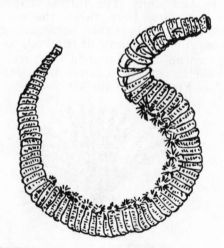

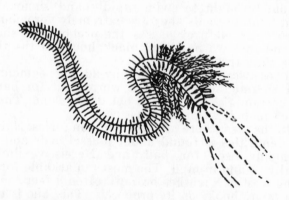

LUG WORM   *Arenicola marina* (*Linn*)

Six to 12 inches long and fat.

Color is green-blue with brownish red tufted gills.

The head is somewhat developed. There are no tentacles, but distinct eyes. There is an exposed sticky, bulbous proboscis with small filelike tentacles and hooks on it. The fore part of the body has hair-like setae on top of the segments, and hooks underneath. Arenicola constructs a U-shaped burrow. At low tide it shows mud castings, for arenicola is always tasting the sub-stratum for food.

This is the worm hunted by fishermen for bait.

Common on our east coast and in Europe.

PLUMED WORM   *Diopatra cupraea* (*Clapareda*)

Average length one inch.

Iridescent with bright red gills like tiny trees. The body is reddish brown flecked with green.

This worm constructs a tough parchment tube that is three feet down. The upper two or three inches extend above the surface like a chimney, and are camouflaged with shell scraps, bits of eel grass, and seaweed. The inside of the tube is smooth and wide enough for the worm to turn around. It sits hanging half way out its door poking its proboscis about to grab passers-by.

Found on the east coast in shallow waters.

The west coast form—Eudistylia polymorpha — is a large worm in a gray or yellow tube 18 inches long. The worm spreads lovely delicate gills out into the water.

Found in tide pools from San Pedro to Alaska.

Sea Worms

ANNELIDS

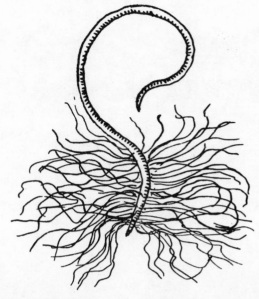

TRUMPET WORM — *Cistenides gouldi* (*Verrill*)

Also known as mason worms.

Average about 2 inches long.

The body is flesh colored mottled with red and blue.

On the head are two sets of golden setae closely set to serve as diggers. These are flanked by bright red branched gills. Though the body shows no distinct segments it is set with parapodia bearing golden setae.

The trumpet worm builds a delicately perfect trumpet-shaped tube of a single layer of finely graded sand grains. In the south it burrows in the mud in mangrove areas, making a lopsided V-shaped burrow with a chamber at the base. As the sand drops into the straight shaft it is sifted for food and shot out the other side.

Abundant from Maine to the Carolinas, but rarer farther south.

FRINGED WORM — *Cirratulus grandis* (*Verrill*)

Often 6 inches long and ¼ inch through.

The color is yellow shading to greenish or orange, or orange-brown.

The general effect of this coloring is iridescence. The cirri or filaments on the parapodia are yellow. There is a red medial line down the body.

The head is small and pointed at the front with no distinct eye marks. The body is round at the head end and flattens out toward the rear. There are no appendages on the first 7 rings, the next two have small bundles of branched cirri. The parapdoia are small.

Fringed worms construct soft tubes in the mud or gravel and beneath stones on our Atlantic coast.

Cirriformia luxuriosa on the west coast lives in the foulest mud with its gills out.

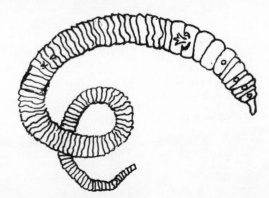

SMALL HEADED WORM — *Capitella capitat*

Average 5 inches.

Bright red. Seven segments have golden bristles. Beginning with the ninth segment there are hooks instead of bristles.

Found in wet gravelly places in all waters, at low tide mark.

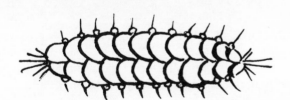

SCALE WORMS — *Lepidonatus squamatus* (*Linn*)

Small worms ranging from one to two inches long.

Yellow to brown.

The body is scaled above and segmented below. Different forms are to be found from Alaska to California on the west coast as well as the east coast.

One form is to be found in the top of the shell of the hermit crabs. These glow in the dark. Those inhabiting the shells of hermit crabs, living in moon shells, have no place in the shell proper, so they climb into the open umbilicus. Many are parasitic, many more free swimming.

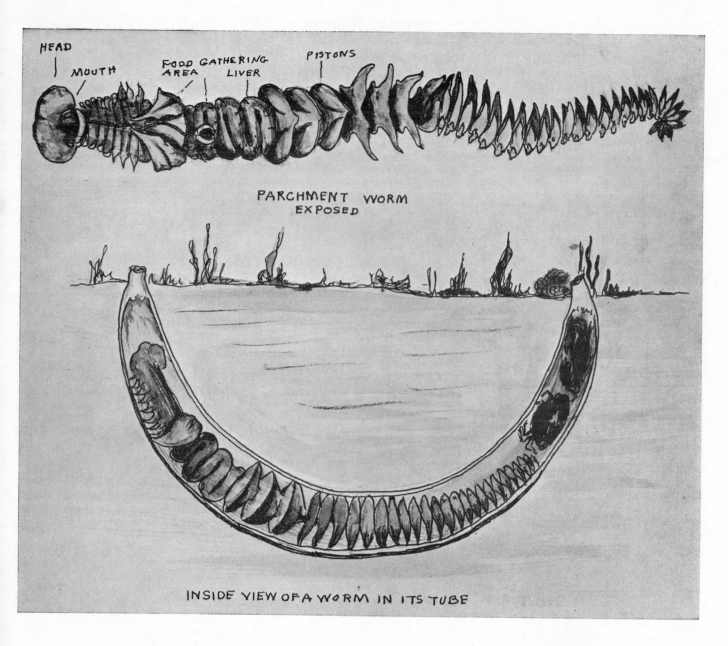

HEAD
MOUTH
FOOD GATHERING AREA
LIVER
PISTONS

PARCHMENT WORM
EXPOSED

INSIDE VIEW OF A WORM IN ITS TUBE

PARCHMENT WORM *Chaetopterus pergamentaceus* (Curvier)

Six inches average. Large 8 inches.

Rather colorless except for the yellow egg coils in the female, and deep green organs just below the head.

These worms live in U-shaped parchment tubes with chimneys showing above the sand and mud. Part way down the body are three pistons that force the water to enter at the head end and flow out at the other. In order not to pump itself out, there are anchors to hold it in place. Microscopic food is gathered on special organs and moved forward to the mouth. While in the larval form they eat diatoms in the sand troughs.

The sexes are separate. Eggs and sperm are ejected into the water where fertilization takes place.

Each year thousands of empty tubes are washed ashore, for there is a great time of dying in the late summer.

The two commensal crabs that live in the tube with chaetopterus are often found in the empty tubes.

Common on the entire east coast.

C. variopedatus on the west coast constructs a woody brown tube. It may be from 6 to 15 inches long. Much slime on animal and in tube seems to contribute to its luminescence. No crabs are found in the tube in the west coast.

Common along the entire west coast.

36

Arthropods

Arthropods are animals with segmented bodies and jointed legs which are adapted for walking, swimming or crawling. Usually the body is in three parts — head, thorax and abdomen. They have special organs for sight, touch, taste, hearing and sense of smell. Arthropods are highly specialized and of wide distribution — they belong on the land, in the sea, and in the air. There are five time as many species of arthropods as in all other species together. In marine arthropods there are a few insects and arachnoids, but the crustaceans, a group of animals with very crusty exteriors, called exoskeletons, make up the vast horde of arthropods living in the sea.

Crustaceans breathe with gills or branchia and have two sets of antennae.

BARNACLES

Barnacles are crustaceans. They begin life as all crustaceans do—free swimming creatures that are most unlike the adults, and as compared to other crustaceans very modified.

As infants they are not at all like the crusty creatures of pilings and boat bottoms. The newly hatched barnacle swims on its back. It has one large eye in the middle of its forehead, six legs on which it never walks, and a single bivalve shell. As it molts two more eyes come into being, and six more legs are added.

Then comes the time the barnacle must change from babyhood to adult. Overcome by an impelling urge it finds a smooth suitable spot and sits down on its head. "Glue pots" that have been developing at its anterior let their powerful adhesive free and the barnacle is set for life. It grows whatever kind of a barnacle shell it is to grow and goes on swimming, but it doesn't go anywhere, for it is stuck by its head. Its delicate body does not rest against the shell — it is wrapped round with a soft mantle. And so it spends its life, breathing with the hairs on its legs, and forever reaching out into the tide with its built-in casting net like a pale hand beckoning the unwary.

Once it sits down and grows its shell the barnacle looses its eyes for it isn't going anywhere and will have no need for them. Nature is stingy with "ocean-probing" eyes. It is presumed that the eyes are absorbed — they are not wasted.

Barnacles are both sexes in one, however, they do not fertilize their own eggs.

There are barnacles that live the quiet life of tide pools, some that live where the only water touching them is splashed there by breaking waves, others ride whales and turtles and boat bottoms. There are stay-at-homes and wanderers, giants and midgets.

GOOSENECK BARNACLE
Lepas fascicularis (*Ellis & Sol.*)

Shell 1 to 2 inches long.

Colorless.

The stalk is shorter than the shell. Paper thin shells are extremely brittle.

Found on seaweeds that bunch heavily, boat bottoms, and any floating object.

Cosmopolitan.

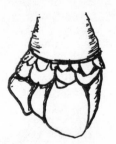

PACIFIC GOOSE BARNACLE
Metella polymerus (*Sowerby*)

Often 2⅓ inches.

In open areas chalky white when dry — in sheltered caves colors are brilliant. The shell is made up of 180 pieces, the whorls decreasing from top to bottom.

Found in association with mussels.

Bering Strait to Lower California. Common.

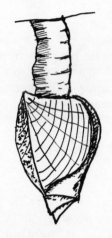

GOOSENECK BARNACLE   *Lepas anatifera* (*Linn*)

Shell 1 to 2 inches long — including neck.

Shells bluish white with scarlet margins. Stalk pinkish brown.

This is the commonest of all ship-bottom barnacles, found in all ranges, more southerly, but world wide.

GOOSENECK BARNACLE   *Lepas hillii* (*Leach*)

Two inches or less.

Bluish white.

Valves smooth, no radial striations on shells. Very long peduncle. The large right shell has an internal tooth at the hinge, but none on the left.

All along the N. E. American coast including Florida.

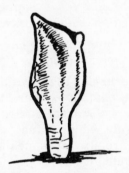

NUDE BARNACLE   *Conchoderma virgatum* (*Spengler*)

May be 2 inches including stalk.

Bluish gray with 6 dark longitudinal lines.

The stalk is flattened, widening out toward the shell. The tip is squarish. The 6 body plates are thin and obscured by membrane.

All ranges — world wide.

Attaches to anything floating—ships, fish or turtles.

EARED BARNACLE   *Conchoderma auritum* (*Linn*)

Length including stalk about 2 inches.

Grayish.

This barnacle has no plates, but is protected by a leathery skin. The stalk is as broad as the body. It is characterized by two ear-like lobes at the upper end of the body.

Brought into all harbors where boats from tropical seas drop anchor. May often travel on whales.

Cosmopolitan.

SOLITARY BARNACLE   *Tetraclita rubescens* (*Darwin*)

About 2 inches across base.
Dull brick red.
This is a solitary barnacle often associated with balanus. The shell is deep ribbed rough as lava slag, very porous, quite conical.
South to middle California.

Arthropods
BARNACLES

GREAT SHIP BARNACLE
Balanus tintinnabulum (*Linn*)

A large barnacle often 2⅓ inches across base.

Shell pinkish red or bluish with white lines. Lip vivid color matching shell.

Since it travels on ships it is common over the entire world. In some countries it is considered good food.

ACORN BARNACLE
Balanus crenatus (*Brug*)

One and one half inches across.

Whitish.

This is a large conical barnacle with rough sharp toothed edges. The plates of the wall are separate from one another. The base is calcareous and very thin.

These are found on stones and shells below low tide.

Common along the North Atlantic coast.

ACORN BARNACLE, ROCK BARNACLE
Balanus balanoides (*Linn*)

Average ½ inch across base.

Whitish to pale brown.

This barnacle has membranous base instead of calcareous. It attaches to pilings, rocks, boats, and occasionally to sea turtles and whales. On the east coast they are on rocks between the tides. They are the commonest and most abundant of barnacles.

Found from the Arctic down.

ACORN BARNACLE
Balanus glandula (*Darwin*)

One half inch across base and ⅓ inch high.

Dirty white.

Most common single barnacle on the west coast. It is found over all the world in all oceans at all depths in stagnant or wild waters. It is so unspecialized that it can adapt to all conditions. The ribs on the plates are deep, the base calcareous.

Found from the Aleutian Islands to southern California.

In intertidal waters they molt most regularly. The cast skins float away in incalculable numbers like small fragile shrimp.

ACORN BARNACLE
Balanus balanus (*Linn*)

Basal diameter often more than an inch.

Dirty white or yellowish.

This is a large conical barnacle with the plates well consolidated. The plates have square edged longitudinal lines. The base is calcareous.

Found attached to scallops and solid objects well off shore.

Abundant north of Cape Cod.

IVORY BARNACLE
Balanus eburneus (*Gould*)

Large barnacles nearly an inch across.
Creamy white.
The shell is low and broad and very smooth. This barnacle has a calcareous base. It is found associated with other barnacles below low tide in salt, or brack, or even fresh water.

Boston to West Indies.

Arthropods
MALACOSTRACA

Lobsters, crayfish and crabs and all higher crustaceans belong in this group. Over 15,000 species are contained in this sub-class. To this group belong the Leptostraca. Animals with eight abdominal segments and a head encased in a sort of bivalve shell or carapace. They breathe with leaf-like gills. Most of these strange creatures are in shallow water among seaweeds — a few are from deep water. To this group belongs Nebalis — as weird a creature as any modernistic mind could conjure up.

NEBALIA   *Nebalia bipes* (*Fabricius*)

Seldom over ½ inch long.

Colorless.

Unlike Nebalia or Mysis, Calanus another of this group is a very graceful, colorful form symmetrically proportioned though small.

This is a small shrimp-like form, slender and laterally compressed. Over the eyes and head is a horny beak-like structure. It has eight pairs of flat legs, the first four pairs are used for breathing and the others for swimming. The last two pairs on the abdomen are vestigal. Fossil forms belonging to phyllocaridans are the forerunners of the malacostraca of the Cambrian, Ordovician, and Silurian periods. The modern forms date from the Carboniferous period.

Nabalia bipes is a filter feeder—setting up a constant flowing stream with its abdominal appendages. The food is sieved out by bristles in the forword appendages. It can also feed on larger particles.

There are no true gills—the eight pairs of leaf-like feet function as gills.

Abundant in shallow weedy waters in all ranges.

OPOSSUM SHRIMP   *Mysis stenolepis* (*S. L. Smith*)

One to 1¼ inches.

Translucent with a black star shaped spot on each segment, and small black spots scattered on other parts of the body.

Has a very decided bend in its back.

Acanthamysis costata of the west coast is but half an·inch long, has very large eyes, more delicate legs. They often drift inshore in hordes.

Found in eel grass from New England southward.

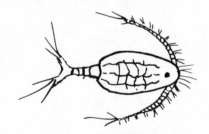

COPEPOD   *Calanus finmarchichus* (*Gunness*)

One sixth of an inch long.

Yellowish, reddish, colorless.

This is the most abundant pelagic form in the northern range. They occur in such great numbers that they color the water.

They swarm so densely off shore that a tow-net will capture many. However, towing is most satisfactorily done at night for great numbers rise to the surface from depths where they have spent the sunlit hours.

Most species have a single median eye. Swimming is jerky. Many copepods are parasitic with sucking mouths, but in this one the mouth is adopted for biting.

Calanus finmarchichus is food for mackerel and herrings, and the most important food of the Greenland whale.

Arthropods

ISOPOD

Isopods are quite bug-like in appearance being flattened from top to bottom for crawling. They are usually slow with weak crawling legs. The thorax has seven free moving joints with seven pairs of jointed legs. There are six abdominal appendages that are used for breathing and swimming. The head bears two pairs of antennae and two eyes. The female carries her eggs in a brood pouch under the thorax. The young are like the parents—there is no metamorphosis. They are the most widely distributed of all arthropods, having 3000 species, mostly marine.

Most isopods are scavengers feeding on dead marine animals. They live under stones and sea weeds. A few are parasitic on fish.

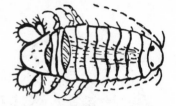

CALIFORNIA PILL BUG
Cirolana harfordi (*Lockington*)

Also known as a sow bug.

Three fourths of an inch long.

Drab.

These crawling pill bugs are the commonest inhabitants of the coast line from British Columbia southward. They are found under rocks high in the intertidal zone. Occasionally they are in mussel beds and curled in dead barnacle shells.

Sow bugs can give a stinging bite that is annoying but harmless.

These are related to the familiar pill bugs found on land under stones and loose bark on fallen trees.

British Columbia on south.

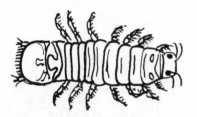

GRIBBLE
Limnoria lignorum (*Rathke*)

One fifth of an inch long and half as wide.

Light gray.

Posterior broadly rounded, legs small for crawling.

The grible is the great destroyer—their numbers are legion and all their food is derived from submerged wood which contains little substance. Needing a constant supply of fresh water they do not burrow deep and are continually making new burrows. Male and female occupy the same burrow, he brings up the rear allowing her to do the work.

Gribbles drill half inch long holes in pilings and submerged timbers, doing great damage to docks. All their nourishment is taken from wood, the food content is low and they do not filter feed.

Range from Florida to Labrador on the east coast, and along the entire west coast.

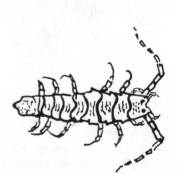

LARGE PILL BUG
Idothea urotoma (*Stimpson*)

Three fourths of an inch long.

Brown.

Identified by its paddle-shaped tail. Slow crawling.

Found on eel grass and under rocks.

Southern California.

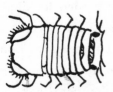

PILL BUGS
Sphaeroma quadritenlatum (*Say*)

One third of an inch long.

Dark and variable. Slaty gray with creamy patches or pinkish edged with black.

Head very broad and short with eyes on outer margin. The first pair of antennae fit into a groove on the underside of the head.

Rolls into a ball when disturbed.

Found on seaweeds, in old barnacle shells, and under stones between tides.

Florida to Cape Cod.

Arthropods

AMPHIPODS

The amphipods have no hard carapace. The body of the amphopod is compressed from side to side. Often the head and thorax are fused. The eyes are not on movable stalks.

There are six free thoracic jointed segments. The abdomen has six segments plus a final segment which forms a telson.

The breathing organs are on the insides of the first joints of the thoracic legs. The first three pairs of abdominal legs are fashioned for swimming; the last three pairs are stiffened for jumping.

The color varies—crimson, brownish red, varying shades of green or gray.

There are amphipods in the depths, in open sea and right up onto the shore where they are learning to live on the land. Disturb a windrow of seaweed left by high tide and they take off in all directions in reckless haste.

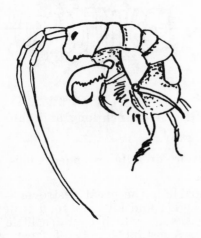

BEACH FLEA   *Orchestoides californiana* (*Brandt*)

Also called sand flea.

Largest of this species. Overall 2½ inches, body length 1 inch.

It is rather a nice looking creature with an ivory white body and bright orange head and antennae. The curve of the body, and the antennae rising high on the head give them an odd rabbity or grasshoppery look.

Like all other beach fleas they are out from dusk till dawn. They use their antennae as grasshoppers do. Like a horde of grasshoppers they rise in great windrows. Beach fleas want to be moist but not wet so they keep just ahead of the tide. During the day they bury themselves in the sand.

Native to California.

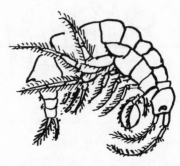

HAIRY BEACH FLEA

Carinogrammarus mucronatus (*Say*)

Average ¼ inch, largest 2/3 inch.

Greenish with minute black, red, or brown spots on the first four abdominal segments.

Antennae are equal, eyes kidney shaped, and the last three abdominal segments have bunches of hair on them. The body is arched and hairy. They are an important fish food.

Found in brackish tide pools and salt water. Cape Cod to Florida.

COMMON BEACH FLEA   *Orchestia agilis* (*Smith*)

Also called sand flea.
One half inch long.
Color variable but mostly brown or olive green. Posterior often bluish, antennae red. Large eyes. Short antennae.

Found on the shores of bays in all areas. They are abundant under seawrack, seaweeds, eel grass, etc. When debris at the tide edge is disturbed they are off in all directions to find new hiding places. They feed on decaying matter.

Mainly they feed on decaying seaweed but will eat any animal matter obtainable. They are easily found but difficult to catch.

Atlantic coast.

Arthropods

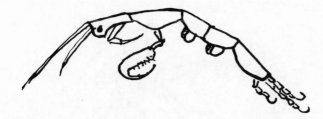

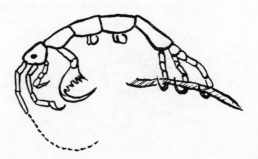

SKELETON SHRIMP   *Caprella kennerlyyi* (*Stimpson*)

Average, mature, a little over an inch long.

Banded with pink. There are two small forward pointing spines on the head.

Watching caprellae under a glass they bow with a solemn dignity, or sway from side to side. They scrape accumulated debris from hydroids and even eat the hydroids. If it becomes desirable to move about they inch along like inch-worms. Often they are present on hydroids in such vast numbers that they form a seething mass.

Alaska to Santa Barbara.

SKELETON SHRIMP   *Aeginella longicornis* (*Kroyer*)

Average size one inch.

Color variable.

Skeleton shrimp are wierd horrors in slow motion, blowing solemnly with cruel sickle-like hands clasped as if in prayer. At rest they are as invisible as walking-sticks, but they can move rapidly inch-worm wise over hydroids where they are so commonly found.

Swishing bunches of hydroids, delicate seaweeds and even eel grass in fresh water will loosen them and they can be strained out. Though they are called shrimp they are closely related to the sand hoppers. They may have spines, tubercles, or be smooth.

The first pair of antennae are twice as long as the other. A very strong hook-like tip on the big hand — as in all other caprellae this is used to snare a passing meal. These animals resemble the land walking-sticks, and are so well camouflaged that it is difficult to see them unless they move.

Found on hydroids and seaweeds.

Labrador to New Jersey.

SKELETON SHRIMP   *Caprella geometrica* (*Say*)

Average 3/5 inch.

Color adaptable to surroundings, colorless or reddish.

Caprellae are queer little animals that are found clinging to seaweeds, hydroids, starfish, eel grass, and sponges. They cling with their strange little walking legs that are at the very end of the body, reaching out at an angle they remain perfectly motionless waiting a chance prey.

This one has no spines or tubercles in its body. The antennae are almost equal.

Very common from Cape Cod south.

Arthropods
DECAPODS

Shrimp belong to the division of decapods, creatures with ten legs. They are crustaceans with compressed bodies a little like flattened cylinders. At the fore part is a rostrum or beak and at the rear there is a strong swimming fin. The first antennae are very long. Swimming is accomplished backward with the tail fin, but forward with the swimmerettes under the abdomen. The female carries the eggs under her tail. Lobsters, crayfish and crabs are decapods.

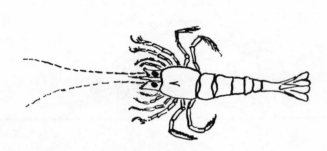

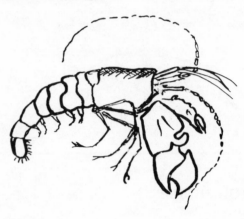

COMMON SAND SHRIMP   *Crangon vulgaris (Fab.)*

Average length 2 inches.

Translucent pale gray, peppered with minute dark spots that make it almost invisible on the sand.

The antennae are almost as long as the body. Behind the small flattened rostrum on the carapace is a spine. They produce abundantly and furnish considerable of the food of the weakfish, kingfish, bluefish and bass.

Found on sandy flats and in tide pools.

Labrador south.

SNAPPING SHRIMP   *Crangon heterochelis (Say)*

Also called pistol shrimp, pistol crab.

This is one of a family of strange crabs ranging from ½ inch to 3 inches in length. This one is grayish, others are clear or pinkish.

The main distinguishing characteristic of these shrimp is an enormous claw on one side. This is supported by a ridiculously slender arm. When annoyed, the wrist of the huge hand is flicked to make a sharp snapping sound. This will stun small prey or repel an enemy by sound — offense and defense by concussion. Snapping shrimp may be found around oyster beds or in mud burrows. One particular shrimp is almost always commensal, living in the large passages of sponges. This one has a bright red snapping claw. They are common to both coasts, each to its own environment.

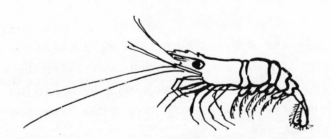

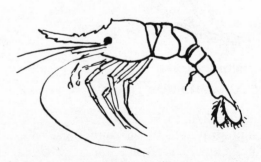

GLASS PRAWN   *Palaemonetes vulgaris (Say)*

Average one inch. Large sizes 1¾ inches.

Colorless — nearly transparent with red spots and lines.

The rostrum is long and straight with 8 to 9 teeth above and 4 beneath. The body is so clear that the internal organs can be seen in a cross light. They swarm in the shallows in eel grass, and rock weed, and in muddy brackish water.

They are in all ranges, but most abundant in tropical waters. These are the shrimp used to feed captive seahorses.

MONTAGUE'S SHRIMP   *Pandalus montagui (Leach)*

Small broken-back shrimp.
Color variable.
Very smooth carapace. Forward half of rostrum toothless above. Abundant in fairly deep water as well as tidal zones.
Found most abundantly north of Cape Cod, but also in the Gulf of Mexico in eel grass.

44

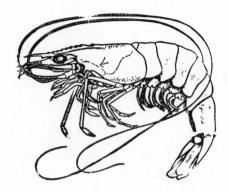

EDIBLE SHRIMP
Peneus setiferus (*Linn*)

Three and one half to eight inches.

Bluish white. Almost translucent.

The protruded beak or rostrum is continued down its back in a low keel, the forward part being grooved. The antennae are very long. The body curved.

They are valuable commercially. Almost 85% of all the shrimp harvested from the Gulf are these. The small ones are called shrimp, the very large ones — prawns.

Found in shallow coastal waters from Virginia to Florida including the Gulf of Mexico.

BROKEN BACK SHRIMP
Hippolyte zostericol (*Smith*)

Average 1 inch long.

Translucent green with red and brown specks, and a median band of dark brown.

The rostrum is straight and as long as the carapace. Two or three teeth on the top of the rostrum and 2 to 4 on the ventral side. There are three spines on the front of the otherwise smooth carapace, and a sharp bend in the abdomen.

They are snappy jumpers living in eel grass below low tide level. Very common.

Related to H. pleuracantha that is found farther south.

Vineyard Sound and south.

RED GHOST SHRIMP
Callianassa californiensis (*Dana*)

Length 2 inches.

Pink and white or rose red.

This is a very clean shrimp to live as it does in soft mud burrowing deeply under rocks. Its legs are admirably adapted to the job — they are made for bracing or holding, and for walking and cleaning. Eating is never ending work, for it eats as an earthworm does. Then there is the cleanup after eating as the used mud must be carried out, and the body must be rubbed and polished— it is a little like the cat that washes its face after eating. It will die shortly if its body is not touching the burrow sides. Often a companion scale worm lives in the burrow with the ghost shrimp.

Common from Alaska to Mexico on the west coast.

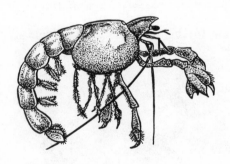

GHOST SHRIMP
Gebia affinis (*Say*)
Upogemia affinis (*Leach*)

Also called mud lobster.

Two to three inches long.

Color — whitish yellow or brown-red.

The body covering is soft and membranous. The first pair of legs have pincer claws of unequal size. The entire body is pubescent, even the antennae. First antenna are small, the second very long.

Ghost shrimp burrow in the mud between tides. The holes are round, running obliquely for about 2 feet, then going straight down or angling off laterally. Unless the bodies of these shrimp touch the sides of their burrows they will die. Food is sifted from mud.

Though ghost shrimp live in the mud they are immaculately groomed, for they have long brushes for cleaning themselves.

Long Island Sound to South Carolina.

Arthropods
LOBSTERS

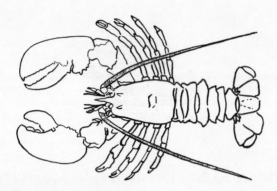

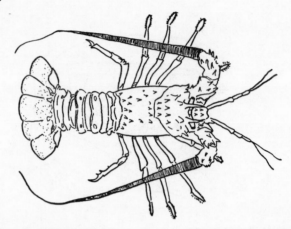

AMERICAN LOBSTER
Homarus americanus (*Miln-Edwards*)

Average size 10 inches; largest ever taken 34 inches; excellent food value.

Color varies, dark brown, dark green to black, usually yellow or orange underneath. Sometimes bright blue on the legs with reddish markings.

The average weight of a marketable lobster is between one and two pounds, mostly over one and a half pounds. This shellfish is the most important food resource taken from along the New England coast. Annual catches are in the millions of pounds. The size and taking is strictly regulated by law.

Found mostly on rock shores, caught in traps baited with fish, Rosefish, one of the most popular baits.

Native to the Atlantic shores from Canada to Virginia.

SPINY LOBSTER
Panilurus argus (*Lat*)

Also called Florida crawfish.

Average size 8 inches; largest 18 inches.

Has a variegated pattern. Usually dull red with decided blue and yellow markings; these can dull to meet the surroundings.

This lobster is entirely defenseless except for its heavy plate armor and the wickedly stout sharp spines. Those in the tail are curved and long. The two long feelers are sense organs as well as being well spiked whips. They will eat most anything. Most of the lobster's life is spent in coral or rock crevices, though they are often found in rocky tidepools in shallow water.

Spiny lobsters are a most important shellfish in the southern waters, not only to man, but to the octopus.

From North Carolina to the Florida Keys and Mexico.

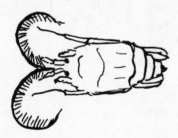

SAND BUG
Hippa talpoida (*Say*)

Also called mole shrimp and sand flea.

Average length one inch.

White tinged with lavender above and yellowish below.

The first pair of antennae are small, but the second are long and plume-like. Tiny eyes are on long stalks. The first pair of legs point forward; the next three pairs are strong with leaf-like extensions on the sides; the last pair are thread-like. The abdomen is bent under — it is long and slender.

Sand bugs live at the edge of the waves, and they keep there as the tide ebbs and rises, sweeping the edges of the waves with their long plumy antennae like cast-netters.

Cape Cod to Florida.

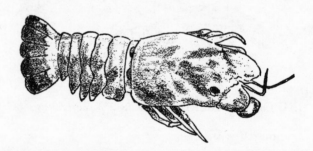

QUEEN CRAWFISH
Scyllarides aquinectalis

Average size 10 inches; largest 18 inches; edible.

Grayish brown on back, light yellow to pink on belly.

A deep-sea lobster without claws. Large, bulky and sluggish. They have seven pairs of legs although a number have been found with only five pairs.

Usually found in deep water, mostly in the West Indies, although occasionally along the Florida coast.

Native to the West Indies, as far north as Key West.

Arthropods
CRABS

PORCELAIN CRAB   *Petrolisthes cinctipes* (*Randall*)

Width - 9/16 of an inch.
Drab.
The very flat bodies of these crabs are well adapted for life under flat stones and in mussel beds. These are the crabs that scurry so madly for cover when a rock is upturned. Vanishing into crevices, and avoiding capture by autotomizing, or casting away, a claw or leg.
Common from British Columbia to Gulf of California.

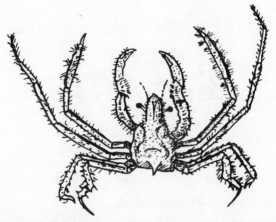

SPIDER CRAB   *Podochela riisei* (*Stimpson*)

Less than an inch long.
Body banded with brown.
This is an ugly little crab with a rostrum forming a thin hood over its head. The legs are much longer in old individuals than in young ones. The body and upper parts of the appendages are covered with tufts of curved hairs — on the lower part of the legs the hairs are straight and stiff. There is a very sharp tubercle at the rear.
Found on both coasts in shallow water.

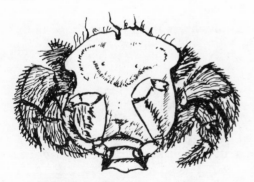

SHELL CARRYING CRAB
Hypoconcha arcuata (*Stimpson*)

A tiny crab scarce ½ inch across.
Color nondescript.
Instead of a hard carapace protecting its back, this little crab is covered above with a parchment-like skin. The underside and the hands are rough and granular. Three pairs of walking legs are developed as such, but the last pair are reduced and held above its back. With these legs this bit of a crab hoists half of a small bivalve shell on its back, tucks its angular rear into the hinge side and carries the shell for its protection. Walking along it resembles a wave washed shell, but when danger threatens, it squats down dropping the protecting shell to the bottom and so goes unnoticed.
Native to the shallow waters from North Carolina to the Gulf of Mexico.

SPONGE CRAB   *Dromia erythropus* (*G. Edwards*)

Average size 3½ inches long, 4 2/3 inches across.
Brown or black stiff hairs on a whitish carapace. Tips of fingers bright red.
This crab may cut a piece of living sponge to fit closely over its back so it may travel unseen by enemy or prey, but usually it sits in a hole cleverly fashioned in the sponge to conform to its body, uses the cut out piece as a lid by clutching it firmly in place with the last two pairs of reduced walking legs which are held dorsally for that purpose.
Brought ashore in nets from North Carolina southward.

Arthropods
CRABS

SPONGE CRAB   *Dromidia antelliensis* (*Stimpson*)

Also called ascidian crab.

Average length 1½ inches, width a trifle under that.

Color is variable. May be dirty yellowish green with carmine pincers tipped with white, or reddish brown with crimson claws and legs the color of horn. Body is entirely covered with short pubescence. The four toothed claws are short and stout and nearly smooth. The rear legs are held above to clutch the covering which is usually larger than the crab. The female's eggs are orange-vermilion.

These crabs are either dredged up or tossed ashore by storms.

Found from North Carolina to Mexico.

FLAME CRAB   *Calappa flammea* (*Herbst*)

Also called shame faced crab and box crab.

Average 4½ inches wide, 3 inches long.

Smoky gray behind fading to drabbish white with stripes and net work of purplish lines. Claws are heliotrope on the outside with purplish markings with some shadings of sulphur yellow; the inner side red. White on underparts. There is a wide expansion on the claws that is topped with a row of teeth. When closed against the body the eyes peer over these teeth reminding one of a naughty child covering his face with his hands and peeping through the fingers. Flame crabs live inshore and out on low corals. When storms arise and things on the ocean bottom are rolled against one another this crab shuts up like a box and rolls about unharmed.

Found in waters from Massachusetts to Brazil.

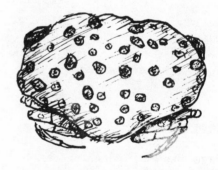

LADY or CALICO CRAB   *Ovalipes ocellatus* (*Herbst*)

Average size two to three inches.

Lavender marked with purple-red spots.

There are five sharp teeth on each frontal margin. The claws are strong and long fingered; the last pair of legs flattened for swimming.

Found on sandy beaches from low water mark to shallow waters, usually lying buried in the sand with just the eyes sticking out.

Good food.

Common from Cape Cod to Gulf of Mexico.

CALICO CRAB   *Hepatus epheliticus* (*Linn*)

Also called Dolly Varden crab.

Average size 2½ inches. Large 3 inches.

This is a crab with a form similar to that of the box crab. The claws are tight folding and high crested. The carapace is yellowish or brownish with dark bordered blood red spots. Big claws of this and the former crab are such that they filter out the grit and sand from the water they suck into their branchial chambers.

Found in all types of bottom except clayey mud. In deep or shallow water.

Chesapeake Bay south.

Arthropods
CRABS

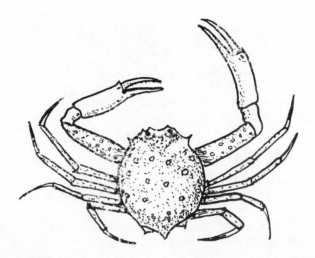

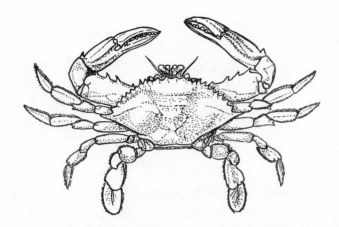

COMMON BLUE CRAB

Collinectes aspidus (*Rathburn*)

Also called soft shelled crab.

Average 7 inches across.

Color is deep olive above, yellowish beneath. The claws are blue with red tipped spines. Both walking and swimming legs are blue.

The blue crab is twice as wide as it is long. The sides of the carapace are drawn out into long strong spiny projections. It has long strong claws. The last pair of legs end in swimming paddles.

This is the highly prized blue crab of commerce.

Found from Cape Cod to Louisiana.

PERSEPHONE

Perseprone punctata (*Rathburn*)

Average size under two inches.

Grayish brown with darker markings.

The carapace is evenly globular with a fine granular surface. Granules are often pink or white, the forward ones being the largest. At the rear of the body are three small, sharp, recurved spines.

Found in shallow water. Not common.

New Jersey to Texas.

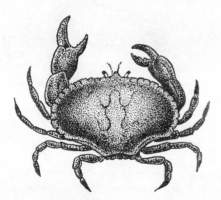

COMMON ROCK CRAB

Cancer irroratus (*Say*)

Average 3 inches long and 5 inches wide.

Yellowish with granular purplish brown spots.

The carapace is convex. Each anterior border has 9 broadly pointed teeth with granular edges. A notch between each two teeth forms a groove onto the body; the last tooth on each side is on a granular ridge. The front has three teeth, the middle one being the longest.

Rock crabs lie buried in the sand with only the eyes out.

Native to all ranges including Mexico. Edible.

THE PURPLE CRAB

Randallia ornata (*Randall*)

Average size two inches or less.

Mottled with purple, brown and dull orange. Claws are variegated red.

The purple crab of the west coast is similar to persephone of the Atlantic Coast, but the tubercles are much smaller.

The legs are rather spidery, but this crab is not closely related to the spider crab. Fine granules cover the round bulbous body. Body divisions are marked by grooves. At the rear are three spines. Purple crabs usually lie half hidden in the sand with their legs curled under them.

Found in sandy bottom on the west coast.

Arthropods
CRABS

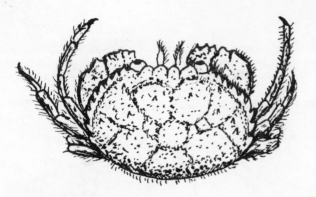

OREGON CANCER CRAB *Cancer oregonensis* (*Dana*)

One and one half inches long and almost two inches across.

Red with a coarse densely granulated surface.

Alaska to Lower California.

NORTHERN OR JONAH CRAB
Cancer borealis (*Stimpson*)

This is very much like the above, both as to size and shape, but the back is brick red with yellowish markings. There is no pointed rostrum.

Found on rocky bottom exposed to surf from north of Cape Cod.

FALSE CRAB *Lithodes maia* (*Linn*)

Average size 5 inch spread; largest 8 inch spread.

Dark brown to almost black with dirty yellow streaks on legs and body.

This is a large spiny crablike form. It is the transition form and really belongs to the hermit crab group rather than to the true crabs. The body is unsymmetrical. At first glance it seems to have but three pairs of walking legs, but the fourth pair is short and carried in the branchial chamber.

Native to the North Atlantic coast.

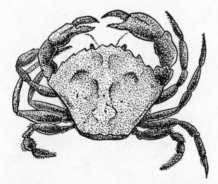

GREEN CRAB *Carcinides maenas* (*Linn*)

From two to three inches across.

Greenish above with light blotches.

Green crabs swim though they have no paddles on the last pair of legs — the legs are flattened with pointed tips. Each frontal margin has five acute teeth. There are molar tubercles on the claws. Green crabs are good fighters and good runners.

Found in the tidal zone hiding under rocks and weeds at low tide, from Maine to New Jersey, We do not consider them good food, but they are hunted for food in Europe.

MUD CRAB *Panopeus herbstii* (*Milne-Edwards*)

Average ⅝ of an inch long, 1 inch wide.

Muddy brown to dark green with black tipped fingers.

The rostral arch has a medial depression. There are five teeth on each lateral side. The legs are adapted for walking.

May be found over rocky or shelly bottom and in oyster beds, hiding under stones at low tide.

Common from Massachusetts to Florida.

Arthropods
CRABS

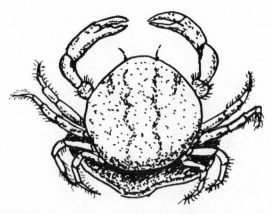

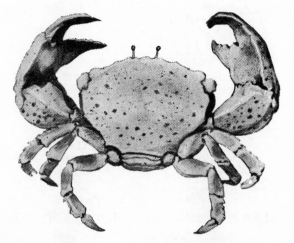

PEA CRAB
Pinnotheres maculatus (*Say*)

Average ½ inch.

Drab in color.

This is not the oyster crab, though it is a relative. The male is free swimming, but the female lives within the shell of a mussel or scallop. Beneath her, her lower shell extends out at the rear giving her the appearance of sitting in a saucer.

The frontal margin is almost rounded; eyes are small; shell weak for she remains protected throughout her life. The male has a crustier shell.

Pea crabs are not parasites. They do no harm other than stealing the hosts food.

Common from Cape Cod southward.

STONE CRAB
Menippe mercenaria (*Say*)

Also known as spotted rock crab.

The average carapace or shell is five inches across. Add to this the length of the extended claws and it is a most formidable appearing crab. However, though equipped with such powerful claws it is slow and amiable.

Color is purplish to brownish red with scattered dark spots. The fingers are black; the legs are banded with red and yellow.

Stone crabs are found in shallow waters and in deep burrows on sandy shoals. There is a closed season on these crabs, and even when open season comes never take the female with the broad tail; take the male with the narrow tail — bring it to shore before removing the claws. Excellent food.

Native to the waters from the Carolinas around to Mexico.

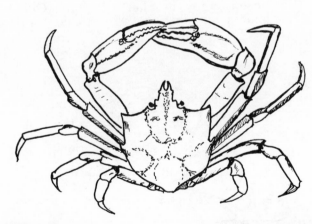

KELP CRAB
Pugettia producta (*Randall*)

A small crab about 2 inches long, and slightly less across.

The top is dark olive green or brownish to match the kelp on which it lives; the under side is light flecked with red.

There are two sharp spines on the sides of the carapace. The spines on the legs are very sharp so the crab can keep a hold in the face of wave shock. The claws are very strong. Large ones are best left alone, but small ones can be handled with care.

Found on kelp from Vancouver Island to Lower California.

ELBOW CRAB
Heterocrypta occidentalis (*Dana*)

Average one inch across carapace.

White tubercles on a pinkish body. Legs usually yellow.

This is the western elbow crab. Characterized by its "elbows" and ridiculously small claws. It walks with its bent arms held out in front of it.

Found near the coast in some areas, but mostly in deep water.

California to Mexico.

FIDDLER CRAB
Uca pugnax (*S. I. Smith*)

Average one inch across.

Light brown with purple and dark brown mottlings.

Fiddlers are belligerent, lively little cowards, seldom seen much below high tide. The males have one claw very much enlarged, giving them a rather ridiculous appearance. It is held horizontally across the front then raised aloft, and waved about, amusingly "resembling the violin section in an orchestra". This claw is of no value in eating, for it reaches far beyond the mouth. If lost in a skirmish, the little claw begins to develop. With the shedding of the skin it is too long for eating, but where the other was lost a small one has developed. The female has no large claw so there is no question as to whether she will eat or not.

Fiddlers live in long slanting burrows in mud or salt marshes. At the end of the burrow is a chamber. This must always be damp but not wet. Fiddlers can be drowned. When the tide is out they travel in droves along the shore, but never far from home. As the tide rises they hurry into their burrows and plug the opening with mud to hold in the air and so hold out the water.

Found on the east coast and the Gulf of Mexico.

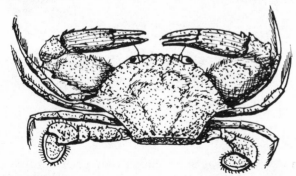

PORTUNUS
Portunus xantusii

Width 2½ inches.

Small relative of the east coast blue crab.

Long side spines are weapons of offense and defense. Extremely active.

West Coast.

GHOST CRAB
Ocypode ablicans (*Bosc*)

Also called sand crab.

Average width 2 inches, length 1¾ inches.

Pale yellow or sand color.

The claws have serrated edges; both are close to a size. Ghost crabs seem to be traveling on their toes as they run. They are the speediest crabs along the beaches. Matching the color of the sand so perfectly, they need only squat to become invisible. They live in burrows high on the beach, but must go into the edge of the water to wet their gills. They must also go to the water to spawn, for it is there the young are hatched and spend their infancy.

Found on sandy beaches along the Gulf of Mexico and as far north on the east coast as New Jersey.

WHARF CRAB
Sesarum cinereum (*Say*)

Also known as wood crab.

Carapace about an inch across.

Tannish brown to black.

Wharf crabs have almost square carapaces. They are very flat and can slip into cracks and crevices on pilings where they live. Unlike most salt water crabs they are found in fresh water, on land, and on the shore. They are a nuisance in harbors for they often collect in droves in the bilges of small boats.

Common from Chesapeake Bay southward.

Arthropods
SPIDER CRABS

Spider crabs do not get their name from the length of leg, but from the sac-like body that resembles that of a spider. Those that live over stony or sandy bottom have short legs; over soft or muddy bottom they are long legged. The eyes and antennae are close together giving the pinched appearance of a spider's eyes.

All other crabs walk or run sideways, but spider crabs can run on the bias.

The largest known crab in the world is a spider crab.

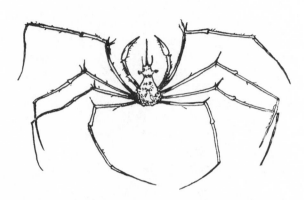

SPIDER CRAB　　　*Metoporhaphis calcaratus* (*Say*)

Average body length ½ inch. The legs are longer in old crabs than in young.

Usually pink or whitish.

The body has a very irregular spiny surface. These crabs are usually found inshore among hydroids and weeds. They are almost invisible. When frightened they extend their legs above their heads like blown umbrellas.

Found from North Carolina to Brazil.

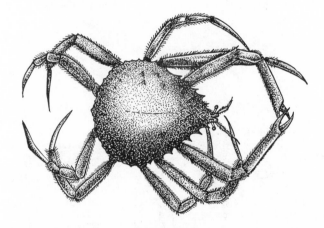

COMMON SPIDER CRAB　　*Libinia emarginata* (*Leach*)

Average three inches long.

Dark brown.

There are nine spines on the body; the legs are long; the body sac shaped. The margin is evenly rounded behind the rostrum. This slow moving crab is often covered with algae, hydroids and barnacles.

Found in shallow water over muddy shelly bottom.

East coast and all ranges.

TOAD CRAB　　　*Hyas coarctatus* (*Leach*)

Average less than two inches, large ones may be 3 inches long.

Dusky grayish brick red above, lighter below.

This is a short legged spider crab with a wide shallow notch in the sides like the body of a violin. The rostrum is split and separated by a narrow slit.

They prefer to live in stony muddy water, and may be found from very shallow water to deep.

Common on the east coast south to Florida and the Gulf.

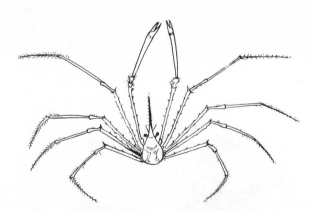

ARROW CRAB　　　*Stenorynchus secticornis* (*Herbst*)

Largest may be 2 inches from the tip of the rostrum to the end of the body. The average is less than half that.

Color is creamy white to buff. Bands of white or chestnut radiate from the center line to the margins of the carapace. The fingers are blue or purple, smooth in juvenile, shaggy in old. The spines on the rostrum, the legs, and teeth in claws may be orange or red.

Common in coral regions from Cape Hatteras to the Gulf of Mexico.

HERMIT CRABS

Hermit crabs seem to have been created in one of Nature's whimsical moods. She gave them armor to the waistline, and then walked out on the job! They are all bravery and bluff in front and cowardice behind. As each infant crab settles down to true hermit crab ways it is very small. It must find a suitable empty shell into which it can tuck its tender rear and pull itself back so as to be completely out of sight. When it retreats within the shell the opening is closed with the two big claws.

Unfortunately its home is a dead shell and so the shell does not grow with the crab, and as the crab sheds its skin and grows it must hunt a larger shell and make a quick change over before some watchful enemy catches it midway.

Hermit crabs are gregarious. They clown around in the shallows like a group of boys in a vacant lot — pushing, tugging, putting on sham battles, and bumping one another over.

COMMON HERMIT CRAB *Pagurus Iongicarpus* (*Say*)
Also caled long armed crab.
Light brownish yellow with striped legs. One variety often found is white.
As in most hermits the first antennae are short, the second long. They are sensory organs. Both hands are elongate and smooth; the right is usually the larger.
They are found most everywhere along the east coast in shallows with either muddy or sandy bottom. They prefer periwinkle or mud snail shells. In northern areas there is a commensal hydroid growing on the shell of this crab.
Maine southward.

LARGE OR WARTY HERMIT CRAB
Pagurus pollicaris (*Say*)
Average mature crab 4 to 6 inches.
May be reddish to bright orange in color.
This is the crab with broad, flat, rugged hands — the right one usually being the larger. The surface of the hands is covered with tubercles. Both claws are used to shut the shell opening.
Found mostly in shells of busycon, natica, and fulgar.
Along the east coast from Maine to Florida, including the Gulf. Most numerous in sounds and bays.

HAIRY HERMIT *Pagurus hirsutiusculus* (*Darwin*)
Average size 2 inches, largest 3 inches.
Body pale, legs banded with blue and white.
A furry body as soft looking as a child's toy. These crabs are seldom found in any other than thais shells. Though they have a wide range they differ materially in size being much larger in colder climates.
Common in San Francisco Bay. Found from Alaska to Lower California.

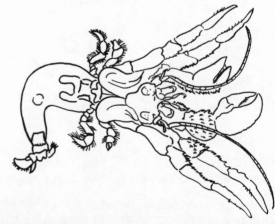

SPINOUS HERMIT CRAB *Pagurus bernhardus* (*Linn*)
Mature form between 2 and 4 inches.
Reddish brown in color.
This is one of the larger hermit crabs, and one of the most common. It has large rounded spinous hands.
Native to the Atlantic and Gulf coasts from Maine to Mexico.

Arthropods

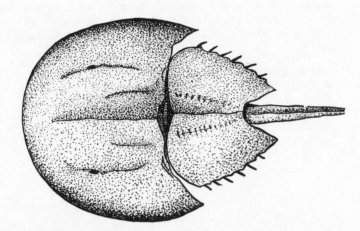

HORSESHOE CRAB   *Limulus polyphemus (Linaecus)*

Average 6 to 14 inches.

Buff to dark brown.

This animal is edible, but the meat is spare.

Horseshoe crabs are a relic of a bygone age, having walked the sea's margins for hundreds of millions of years unchanged. Limulus is not a crab as its common name implies, but the last living remnant of an otherwise extinct race. A crab has a face all in the proper order; the horseshoe crab has four eyes on top and a mouth on the underside — that will not constitute a face. It is the strangest mouth in the world — surrounded by knees instead of lips, and the teeth are not in the mouth but on the knees. As it walks along its knees rub together, so the teeth grind together. Putting small creatures it plows from the mud between its knees it chews them. When it stops walking it stops eating. It must march or mark time to eat.

Beneath the abdomen are a double row of breathing organs called gill books. Each delicate book is covered by a hard plate.

In deep water it swims on its back, a relic of its upside down life in the egg.

There is a tiny pair of pincers ahead of the five pairs of pincer-bearing walking legs: another pair of legs behind these end in sharp spikes surrounded by flanges that open in walking and resemble skiier's poles. These help push the horseshoe crab along its mud-plowing way.

Horseshoe crabs are found only on the east coast of Asia and the east coast of United States including the Gulf of Mexico.

MANTA SHRIMP

Manta shrimp are elongated shrimp that keep their tails held out straight behind them. The gills are under the abdomen or tail. The front legs have terrific claws which are held as a praying mantis holds its claws. In the rear are a set of sharp daggers. They make fine burrows with several entrances a few feet apart. Lying partially out of the burrow the mantis shrimp lays its claws extended all the easier to catch any creature coming near. The nickname "old rip thumb" is very apt. The female carries her eggs in some of the mouth parts instead of under the abdomen.

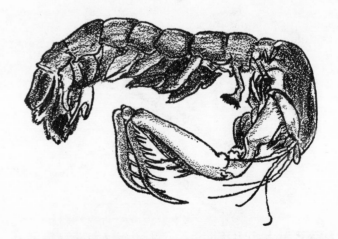

MANTIS SHRIMP   *Squilla empusa (Famricus)*

Average five to seven inches long, however, 11 to 12 inches is no rarity.

Greenish gray with yellow or brown trimmings. The eyes are a beautiful emerald green. This species has 6 teeth on each claw. They are edible.

Found from Cape Cod to Florida and the Gulf of Mexico.

Arthropods

SEA SPIDERS

Sea spiders are curious stilted spider-like creatures found in all oceans at all depths. Most of them are small. They have four pairs of long slender walking legs and a pair added to carry the eggs. Almost the entire body is taken up by the cephalothorax, which is a body-head arrangement. The abdomen is extremely small. Much of the internal economy is carried in the legs.

The male has an extra pair of legs with which he carries the eggs after the female lays them however he is merely a carrier. The newly hatched young have but three pairs of legs.

Sea Spiders have a digestive, nervous and a circulatory system, but no respiratory system. The body has a cephalothorax head and thorax combined, and an amazingly rudimentary abdomen which is insufficient to accommodate all the body organs, so like the starfish, a tubular outgrowth of the legs contains the stomach in these. The egg sacs are in the legs and open at the basal joints.

These creatures are found inshore on algae and hydroids. Of these the smallest are 1/25 of an inch long is the largest ½ to 1½ inches long. Very deep sea forms have a spread of several feet and are bright scarlet. Some small forms living in shore are found on sea squirts and hydroids on pilings.

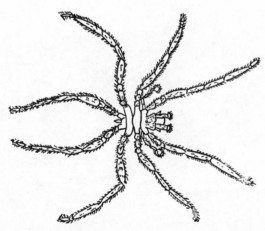

SEA SPIDER   *Nymphon stromii* (*Kroyer*)

Body 3/5 of an inch long with a leg spread of 5½ to 6 inches.

Pinkish yellow. Legs often ringed with red.

This is a very common species found from Labrador to Long Island Sound in from 75 to 600 feet, on rocky or gravel bottom.

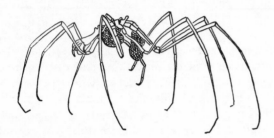

DEEP SEA SPIDER   *Nymphon hirtum* (*Fabricius*)

The body is about 3/5 of an inch long with a leg spread of 4 to 6 inches.

Light salmon-yellow with legs banded with red or light purple.

The slender body is quite smooth. Legs long and sparsley hairy. Very round black eyes.

On rocky bottom between 75 and 600 feet deep.

Common on the Atlantic coast from Long Island to Labrador.

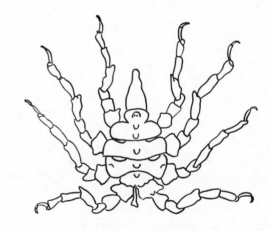

FUNDY SPIDER   *Pycnogonum littorale* (*Grom*)

Average size 2/3 of an inch.

Light yellowish brown to dark brown. Tips of legs black.

This little sea spider is broad and flat with small widely separated black eyes. Each body segment has a conical tubercle.

Usually found under stones and on sea anemones.

Native to the Atlantic coast, but most common in Long Island Sound and the Bay of Fundy.

Echinoderms

Echinoderms are definitely creatures of the sea. They are found in all seas at all depths from the tidal zones to the great deeps. There are five members of the family: starfish, brittle stars, sea urchins, sea cucumbers, and sea lilies, each having five divisions to the body and built on a radial plan. Most all have stalked pincers and movable spines. Tube feet are present in all, but all do not have suction cups. Their mouths are on the ventral side, but the sea lily sits upside down on a stalk, and the sea cucumber travels on its side. Echinoderms are the only creatures in the world that have tube feet manipulated by a water vascular system. They have no relatives on the land or in fresh water.

STARFISH

The starfish is one of the most voracious of predators on the ocean floor. If a starfish wishes to eat a fair sized snail, it avails the snail nothing if it slams its trap door shut — the starfish simply extrudes its stomach through its mouth, wraps it around the snail to smother it. If the prey is a bivalve the starfish squats over it and, with attached tube feet holding fast, pulls the valves slowly apart. In goes the extruded stomach paralyzing the bivalve and the starfish eats it at leisure. No other creature in the sea can extrude its stomach; some starfish cannot, either.

The general pattern for starfish is five points, but there are exceptions. An eye spot, which is sensitive to light, is at the tip of each point or arm.

Points or tube feet lost are replaced. On the top of the central disc is a round spot marked differently from the rest of the disc — this is the sieve or madrepore plate to sieve the water that passes into the water vascular system that works the tube feet.

Some starfish travel on tube feet and some are able to swim. In very troubled seas starfish can bury themselves in the sand or cling tenaciously to rocks.

ASTROPECTEN   *Astropecten articulatus (Say)*

Eight to 10 inches in diameter.

Purplish above, yellow beneath, bordered with orange-red or purple spines.

There is a double border of wide marginal plates — the lower row heavily edged with spines. The small tubercle at each arm tip is the eye spot. There are no suckers on the tube feet, and no pedicellariae or pincers present.

New Jersey to the Gulf of Mexico in shallow water.

ASTROPECTEN   *Astropecten comptus (Verrill)*

May be up to 10 inches across.

Colorless.

Has long tapering, but quite stout arms with acute angles at the disc. The marginal plates are rounded oblongs well raised above the rest of the body. The surface is granular showing a pattern of flowerlike rosettes. There are no pedicellaria. There is no anus and no suckers on the tube feet.

Found from Cape Hatteras around Florida. Shallow water down to 60 to 150 feet.

Echinoderms
STARFISH

WEST INDIAN ASTROPECTEN
Astropecten duplicatus (*Gray*)

Three to 4 inches across.
Rather colorless.
Characterized by sharp angles between arms. Marginal plates large and raised. No suckers on tube feet. The surface is granular with bristling marginal spines. These can paddle along on soft bottom. They are not quick to fragment.
Found from North Carolina to Florida and the West Indies in shallow water to 90 feet.

BREAKING STAR
Luidia elegans (*Perrier*)

Large ones are 14 inches across.
Rich orange above — paler below.
Pattern on top is rosettes. There are no marginal plates. The pincers or pedicellarias are two valved. These wash in after storms. They are usually in deep water.
Found from New Jersey to Cape Hatteras and the West Indies.

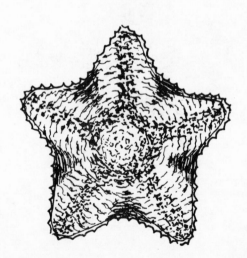

MUD STAR
Ctenodiscus crispatus (*Retzius*)

Three to four inches in diameter.
Color yellowish.
These show great variation in shape. They may be almost pentagonal to regular star-shape. A silken skin covers the large marginal plates. Starfish with broad marginal plates have no anus, but eject leavings through the mouth.
Circumboreal. From Arctic to along the South American coast. From shallow water to 6000 feet.

BREAKING STAR
Luidia clathrata (*Say*)

Four inches in diameter.
Salmony-pink or gray.
The arms are long and slender with rows of mosaic-like tubercles, and no marginal plates above. When buried they are easily located by star-shaped markings in the sand above them. These are depressions made by the starfish taking in water below for breathing and exhaling through the upper surface thus "blowing" away some of the sand. They creep rapidly around with an oar-like movement of their powerful flattened tube feet, but can also swim.
Found on rocky shores in shallow water from South Carolina to Brazil.

Echinoderms
STARFISH

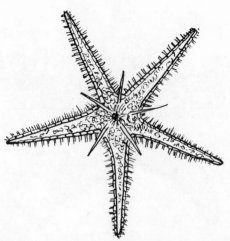

ARMATUS — *Benthopecten armatus* (*Verrill*)

Nondescript in color and of varying size.

Very flexible. Marginal plates on arms are long and slender, each with a long spine. Five very sharp spines radiate out from the mouth area. The tube feet are equipped with quite small sucking discs.

Circumboreal, appearing in our waters from New York to the Gulf of Mexico. A slight variation of form may be found in different waters.

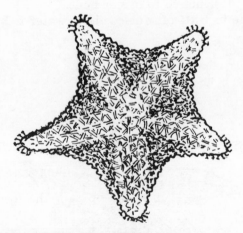

RETICULATED SEA-STAR — *Orester reticulatus* (*Linn*)

Twelve to 16 inches across. Largest 20 inches.
Almost all shades and colors.
This is the largest sea-star on the Atlantic coast. The disc is high. The entire surface marked with a reticulated skeleton that divides the surface into triangles. The whole is covered with a leathery skin making the marginal plates hard to distinguish except that they are covered with blunt conical tubercles.

They are found lying on the sand in shallow water.

Found off Florida toward North Carolina to the West Indies.

BLOOD SEA-STAR

Henricia sanguinolenta (*O. F. Muller*)

One and a half to 4 inches.

Mainly rich red. Also orange, rose, purple, or cream.

The blood stars have small discs with long cylindrical arms. Tube feet are in two series. Spines very small.

Eggs are carried around the mouth in a brood pouch made by curling the body legs around them, and retained until the blood stars are real sea-stars.

Found from tidal zone to deep water in rocky coasts in tide pools.

Greenland to Cape Hatteras. Common in Maine and Massachusetts.

WESTERN BLOOD SEA-STAR — *Henricia hendici*

Five inches.

Blood red with long sharp rays. Related to the H. sangunolenta of the east coast.

Echinoderms
STARFISH

SENTUS   *Echinaster sentus* (*Verrill*)

Four and a half inches across.

Colorless.

Small disc with long pointed arms having 2 rows of marginal plates with wart-like spines. Tube feet in two rows with unusual extension. Sucking discs present.

Reef dwellers in shallow areas from N. Carolina to Florida and the West Indies.

COMMON STARFISH   *Asterias forbesi* (*Desor*)

Six to 11 inches.

Color variable — brown, purple, greenish black, or orange-green.

Four rows of tube feet bearing sucking discs.

Madrepore plate deep orange-yellow. Usually five arms, sometimes 4 to 6 arms. The arms pinch in where they meet the dome-shaped disc. The body is covered with strong interlocking plates that make a mosaic pattern.

Very destructive to oyster beds.

Maine to Gulf of Mexico. Low water to about 160 feet.

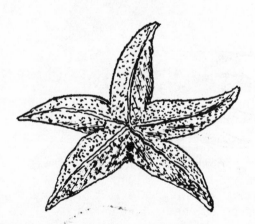

SPINY VIOLET STAR   *Echinaster spinulosu* (*Verrill*)

Four inch span.

Dark violet.

Disc small with long round arms. Covered with a meshwork having long spines on the outer surface.

Tidal zone to shallow offshore water.

Florida to Cape Hatteras.

NORTHERN STARFISH   *Asterias vulgaris* (*Verrill*)

Six to 12 inches. Largest 17 inches.

Variegated yellow-brown and purple. Madrepore plate yellowish orange. Spines lighter than the body.

Arms are flat and pointed. The skeleton very weak, so this starfish must brace its arms in pulling.

Bay of Fundy to Cape Cod.

Echinoderms
STARFISH

WESTERN BREAKING STAR

Linckia columbiae (*Gray*)

Four inches.

Gray and red mottlings.

Seldom is a symmetrical one found. Arms, madrepore plates and mouths vary from one to several. Even when undisturbed it may have a leg twist loose and travel off to grow a new starfish.

Even a piece of a leg minus any part of the disc can regenerate a body.

Found in southern California and southward.

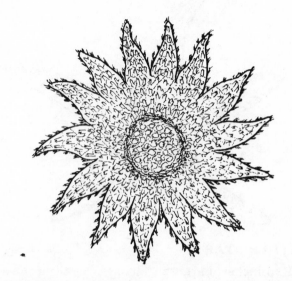

COMMON SUN-STAR   *Crossaster papposus* (*Linn*)

Fourteen inches in diameter. Eight to 14 arms.

Center of disc scarlet blending to crimson. Arms banded with pink or white, tipped with crimson. A pinwheel of brilliant color. Bluish white beneath.

Will eat any creature it can capture even to sea anemones and is cannibalistic.

Circumboreal — Arctic to New Jersey on the east coast and Vancouver on the west coast.

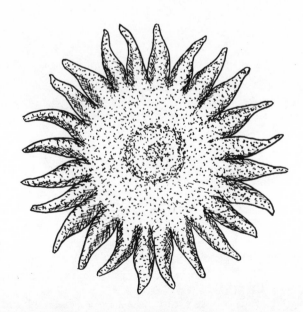

SUNFLOWER OR MANY RAYED STARFISH

Pycnopodia helianthoides (*Brandt*)

Two feet across. Largest of the starfish.

Soft skin is colored delicate pink and purples.

A very active starfish with 21 to 24 rays when mature, though it begins life with six. It adds and sheds arms constantly.

This is the worst enemy of the moon snail.

Found from Unalaska to San Diego.

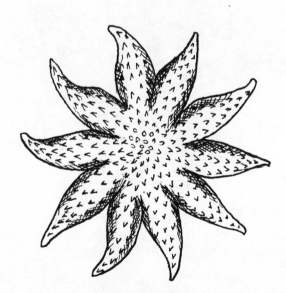

ELEVEN ARMED STARFISH   *Solaster endica* (*Linn*)

Sixteen inches in diameter. Seven to 13 arms. Usually 10.

Red-violet with yellow gold marginal plates and a light yellow madrepore plate well to the side of the disc.

Shallow water from Cape Cod northward.

Echinoderms
STARFISH

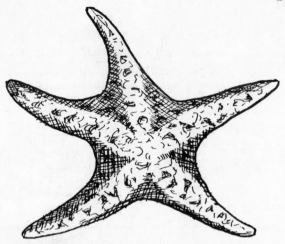

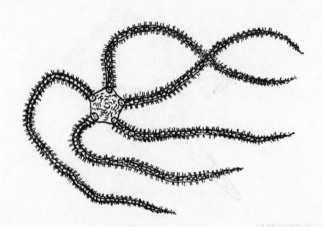

LEATHER STAR *Dermasterias imbricata* (*Grube*)

Two inches. Largest 10 inches. Small in south, large in north.

A soft skin, delicately marked with purple and red conceals the skeleton. The tips of the arms are turned up.

Alaska to Monterey, Cal.

LONG ARMED SNAKE-STAR
Amphipholis squamata (*Delle Chiaje*)

Disc one-third inch across. Arms one and one-half inches.

Grayish white tinged with blue.

In the very young the disc is orange. Both sides of the disc are scaled.

Found on eel grass and at the roots, in dead shells and crevices, from low tide to 60 feet or over.

Arctic to Long Island Sound, and from Alaska to San Diego, California.

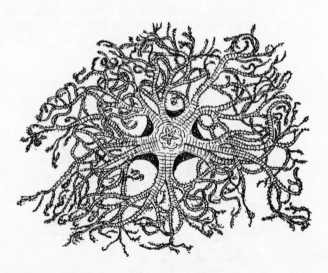

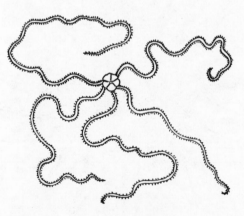

BASKET STAR *Gorgonocephalus arcticus* (*Leach*)

Disc may be from 2½ inches to four inches across. Arms as much as 14 inches in very large ones.

Color varies from cream to dark chocolate brown. May be a shade lighter below.

This star has a very basket-like appearance when its many branched arms are curled up below as it lies on the sand waiting for prey, it walks on its branch tips, and often is found spread out on gorgonians to form a fishing net.

Found on the east coast and the Gulf of Mexico. Moderate to deep water.

LONG ARMED SNAKE-STAR
Amphiolus macilentus (*Verril*)l

Disc one-eighth inch across. Arms two and one half inches long.

Usually light gray.

Disc has overlapping scales like shingles forming a rosette. Arms are very slender and fragile.

Found in soft mud and muddy sand from shallow to deep water.

Cape Cod southward.

Echinoderms
BRITTLE-STARS

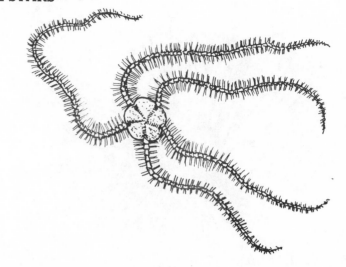

LITTLE SPINY BRITTLE-STAR
Ophiothrix angulata (*Say*)

Disc one-half inch. Arms two and one-half inches.

The color is extremely variable and of no value in identification. The arms differ in color from the disc and are banded. Disc extremely spiny with large bare heart-shaped radial discs. The arm plates are heavily spined.

Inhabiting shallow water and to a depth of 1200 feet.

New Jersey southward.

FRAGILE BRITTLE-STAR   *Amphiodia occidentialis*

With eggs present the disc is swollen to one-half an inch across. The arms are long and delicate with tiny spines set at right angles. It is very difficult to preserve complete, for it sheds its arms bit by bit rather rapidly when picked up.

During the night these stars crawl about, but bury in the daytime in soft substratum rather than under rocks.

Alaska to Monterey Bay.

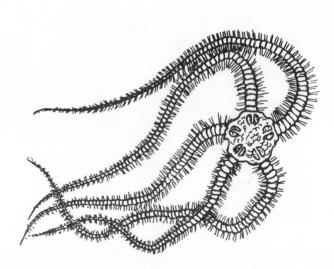

DAISY BRITTLE-STAR   *Ophiopholis aculeata* (*Linn*)

Disc is four-fifths of an inch across. Arms may be from four to five inches long.

Disc often reddish. Arms brown with gray spottings. White below. The central disc is separate from the arms. The slender arms are made up of segments that give them a snake-like appearance, and writhe to propel the brittle-star on the sand or through the water. The arms are edged with spines. The stomach is not eversible.

Most of the feeding is done at night.

Daisy brittle-stars live among seaweeds and corals, over sandy bottom where they submerge for daytime safety, and from shallow water to deep, throughout the entire north temperate zone.

FLORIDA BRITTLE-STAR
Ophiophragmus wurdimani (*Lyman*)

Disc two-fifths of an inch. Arms four to five inches long.

May be cream, dark gray or brown on disc. Arms are ringed with dark gray and cream.

Scales on disc are smooth. The dorsal side is marked with five heart shaped radial plates. The teeth are broad and flat.

Formerly found only around Florida, but now extending as far as the Carolinas.

Echinoderms
SEA-URCHINS AND SAND-DOLLARS

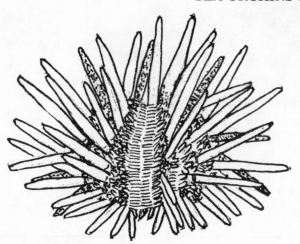

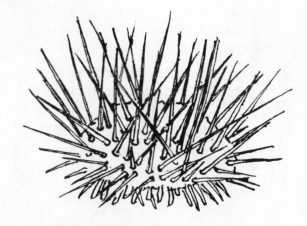

SLATE PENCIL URCHIN
Cidaris tribuloides (*Lamarck*)

Average about two inches.

Light brown shaded with darker brown, dappled with white due to growths; occasionally they may be green, olive or red. At times the spines may be banded with yellow and red.

There are three types of spines — short, sharp, slender ones set in longitudinal rows: which are parted down the middle. Flanking these are flat spines scarcely longer, some surrounding the double row of long heavy spines which are as long as the disc is wide. In earlier times these long spines were used as slate pencils.

This is a tropical species found from Carolinas to Brazil, in shallow water and on reefs. When the tide ebbs those in shallow water hide near low tide. They are solitary.

BLACK SEA-URCHIN
Centrechinus antillarum (*Philippi*)

Average four inches in diameter, two inches high.

Test black or purplish black. Because of the 12 to 15 inch spines it is often called the long spined sea-urchin. Due to a mucous coating the spines are poisonouus to the touch. It is like a hornet sting, and though seldom serious physically, it can certainly upset one's morale. A shadow will cause the spines to point to an intruder; it is a passive attack, but effective.

They are found in shallow water among rocks and corals, often sinking into small hollows and presenting an array of spines to the outside world. They are also associated with loggerhead sponges.

Florida and the West Indies.

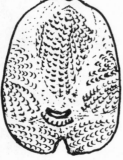

HEART URCHIN
Moira atropos (*Lamarck*)

One inch wide, one and one half inches long. Yellowish.

Egg-shaped. Width and diameter equal. Mouth is well forward. The grooves are deep slits for this urchin lives in soft mud in a slime tube and the grooves allow water to flow over it more freely. Usually only bleached tests are washed ashore.

From North Carolina southward.

ARROWHEAD SAND-DOLLAR
Encope emarginata (*Leske*)

Five inches long, slightly narrower.

Mostly dark maroon.

Solid and stony. Only one of the six marginal holes is completely enclosed by the disc. The others are open at the margin.

Shallow water on the Florida west coast, and the entire Gulf coast.

Echinoderms
SEA-URCHINS

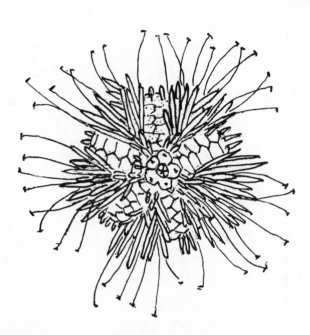

GREEN SEA-URCHIN
Stronglocentrotus drobachiensis (*O. F. Muller*)

Broad and low, averaging two to three and a half inches across.

The close set fluted spines are one half inch long. The test is greenish brown with green spines with no naked area. The tube feet have sucking discs.

The strange fragile teeth are capable of grinding up limy waste and coral rock.

Green urchins are circumpolar and on both coasts, from tide pools to 650 feet and where turtle grass is found.

PURPLE SEA-URCHIN
Arbacia punctulata (*Lamarck*)

One to two inches across.

Purplish brown, with ten rows of very long red or brown tube feet.

On the top or apical pole there are no spines. The rest of the shell is covered with long stiff spines varying in length from one to one and a half inches. This urchin walks rather fast with a queer tilting rolling movement.

They are found on rocky shelly bottom from Cape Cod to the Gulf of Mexico, hiding in crevices and weeds when the tide is low.

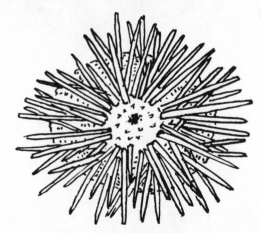

ROCK BORING SEA-URCHIN
Echinometra subangularis (*Leske*)

An eliptical urchin two inches or less in length.

Test may be red-brown to violet or black. Spines, ¾ inch long, are pink or green at tips. There is a great variation in shape and color.

Lives in shallow water on mud flats from Carolinas to Florida.

COMMON SEA-URCHIN
Toxopneustes variegatus (*Conrad*)

Two to three inches in diameter.

Test deep violet, green, or white. Spines short and slender — set in very distinct rows. May be tipped with rose or entirely pink.

This is a decorator, using its suction cupped tube feet to gather debris and shells to coat itself for protection.

Found on sandy bottom from low water to 30 feet, from N. Carolina to West Indies.

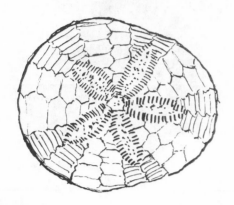

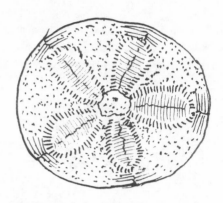

CAKE-URCHIN   *Clypeaster subdepressus* (**Gray**)

Six to seven inches long, five inches wide.

Reddish to brown, in deep water. Yellowish or greenish in tidal waters.

Oval. Raised in center. Very short spines with a petaloid area like the sand-dollar. Found in tidal waters to forty feet, from the Carolinas to Brazil.

CAKE-DOLLAR   *Echinanthus roscaceus* (**Linn**)

Length four to six inches. Width five inches.

Yellowish green to brown or reddish.

Petaloid area depressed.

Sandy bottom in low water, and on open beaches exposed to surf.

Carolinas to Bahamas.

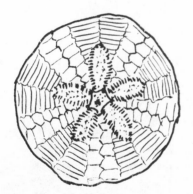

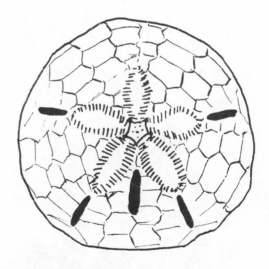

SAND-DOLLAR   *Echinarchinus parma* (*Lamarck*)

Average three inches. Very thin.

Purplish brown spines turn greenish in the air. Spines thick and velvety. The movement of the spines is more like changing light than actual movement. Travel slowly on spines and the tube feet.

Preyed on by all fish with gristly bones, and by flounders, cod and haddock.

Found on sandy bottom at low water mark, often buried.

Circumpolar from Long Island to Puget Sound.

KEYHOLE-URCHIN   *Mellita testudinata* (**Klein**)

Two to five inches.

Pale brown to greenish.

Round disc, flattened at the rear. Mouth below in center. Five slots. Small close set spines. Anus beneath close behind mouth. Iodine smelling yellow fluid secreted when picked up alive is harmless and washes off easily — it is a protective substance to deter enemies.

In shallow water, often buried, from Cape Hatteras to West Indies.

Echinoderms

SEA-CUCUMBERS

Sea-cucumbers, the holothurians, are members of the starfish family. There is a definite radial symmetry that can easily be seen by looking at the mouth. Unlike any other in the group they all have leathery skin. In many, the tube feet are not in lines of fives, but scattered or in irregular arrangement. Breathing is accomplished by the upper tube feet, and traveling by those beneath.

When annoyed they may either eject the entrails or rupture to send them into the water. The enemy stops to eat, and the sea-cucumber gets safely away, to grow new ones.

Most sea-cucumbers earn their keep in the world by cleaning out decaying and small living organisms filling the infinitesimal spaces between the mud and sand particles. In a few places in the world, they are eaten by humans.

SEA-CUCUMBER   *Cucumaria frondosa* (*Gunneus*)

Six to 12 inches long, and three to four inches thick.

Reddish brown above, paler beneath.

Cucumaria frondosa is thick and quite rounded at the anal end. Around the mouth are ten tree-like tentacles about an inch long. The retractile tube feet are in five long distinct lines.

Found from Nantucket north in the lower tide limits.

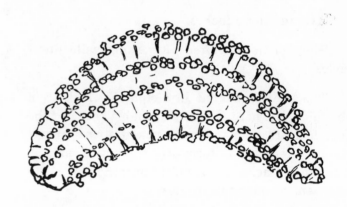

SEA-CUCUMBER   *Cucumaria pulcherima* (*Ayers*)

One and a half to two inches long.

White or pale yellow.

This sea-cucumber is U-shaped — the dorsal side being longer than the ventral. There are five rows of tube feet with suckers, each row having two double lines of feet. This particular one can walk regardless of which side is up. At the mouth end are ten much branched tentacles — the two oral ones shorter than the rest.

After storms, when much eel grass is washed ashore, these are found in quantaties.

They live in shallow water over muddy bottom where eel grass grows.

New Haven southward.

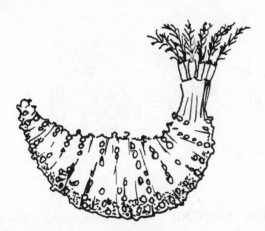

ROUGH-SKINNED THYONE
Thyone scabra (*Verrill*)

Average two inches.

There are a great many varied shaped calcareous plates in the skin, giving the entire body a very rough much folded surface. The tube feet are scattered.

Bay of Fundy to below Delaware Bay from tidal waters to 600 feet.

Echinoderms
SEA-CUCUMBERS

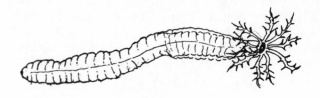

SINGLE STRIPED THYONE
Thyone unisemita (*Stimpson*)

Average three inches.

White or flesh colored with orange-yellow tentacles.

The body is curved in a stout U, thickest in the middle. There is a single double row of tube feet beneath, flanked by a naked strip; above, the tube feet are uniform over the entire surface. The ten tentacles are sparsely branched with the two ventral ones the shortest.

New England southward over sandy bottom in depths from 60 to 250 feet.

COMMON SYNAPTA *Synapta inhaerens* (*O. F. Muller*)

Four to six inches long, ¼ to ⅜ inches wide.

Whitish tinged with yellow or pale red.

This cucumber is as flexible as a worm and quite lovely. The twelve tentacles gather in sand and mud which after being cleared of food is deposited in tiny mounds at the surface. Though they usually lie buried in the sand or clear mud, they are often found exposed, or under stones.

Maine to S. Carolina.

TUFTED SYNAPTA
Chiridota laevis (*Fabricus*)

Four to six inches in length.

Pink to transparent white.

The twelve short stalked tentacles have ten short finger-like branches at the ends, with no tip finger.

Found from low water to 250 feet from the Arctic down both coasts to Cape Cod on the east coast.

COMMON THYONE
Thyone briarcus (*Lesueur*)

Four to five inches in length.

Dull brown, olive or black.

Ten branched tentacles have six branches and a tip. The tube feet are spaced irregularly over the entire body, which is plump sack-like, and thickest in the middle.

Moss Animals and Lamp Shells

These are two of the oldest forms known to man. Fossils dating back to the Cambrian rocks show that these two creatures were well developed hundreds of millions of years ago.

Moss animals, the Bryozoa, are minute marine animals living mainly in colonies, some upright, like mossy seaweed, some incrusting. The upright may range from tiny tufts less than ¾ of an inch high to several inches, from microscopic to those easily seen by the naked eye. The delicate lacey traceries so often found on rocks, shells and seaweeds are incrusting bryozoa. These are often beautifully colored ranging from white through shades of yellow to brick red. Bryozoa live between the tides and to deep water.

At one time the upright ones were thought to be seaweeds, but now we know that each minute receptacle that goes to make up the aggregate contains a microscopic animal. Though it is sessile, it has free moving tentacles and a door it can slam in an emergency.

Two groups frequently seen are Crisia and Membranipora. Crisia is upright, Membranipora incrusting. Both begin as eggs that are developed into larva before being shed into the water, however, the colony grows by budding.

MOSS ANIMALS

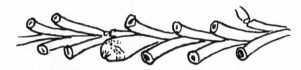

MOSS ANIMAL   *Crisia eburnea* (*Linn*)

Height of colony ½ to ¾ inches.

White with yellow joints.

Form tiny tufts delicate, erect, branching, and dense. The zooecia or cups that enclose the animals are slightly curved and cylindrical. They alternate up the stem in two rows. The larger rather bulbous cups produce the embroys that begin new colonies.   Found on all coasts, world wide.

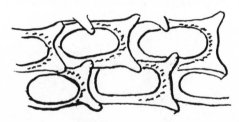

SARGASSUM LACE

Membranipora thuelcha (*d'Orbigny*)

This bryozoan forms a frosty white network on Sargassum Weed. As this weed is torn loose and floats its distribution is world wide. Another form is found on kelp on the west coast.

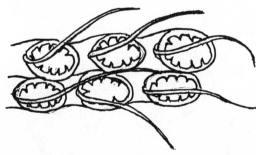

SEA LACE   *Membranipora pilosa* (*L*)

This is an incrusting form that grows around the bases of gorgonians and seaweeds and incrusts rocks and stones. They form long lines on eel grass and large conspicuous patches on stones. There is a long spine arising from the bottom of the opening of each zooecium.

LAMP SHELLS

Lamp shells or Brachipods have bivalve shells, but they are not molluscs. The shells are different size and are without hinges. Molluscs have left and right valves — lamp shells have ventral and dorsal shells. There may be a hole in one valve through which a muscular peduncle or stalk extends. This is used either for attachment to a firm surface or to help anchor the animal if it burrows and lives in a tube in the sand. By relaxing the stalk the animal is raised in its burrow to the surface to feed, by kinking, the lamp shell is drawn down into the burrow.

This is the oldest known animal genus, dating back 500 million years.

LAMP SHELL   *Crania anomala* (*O. F. Muller*)

Crania has no stalk, but is attached to some surfce so closely that the ventral valve conforms to that surface. It has a squarish shape, is white, and the animal within is white tinged with yellow or brown. Stiff hairs or cirri extend out the edge to direct water currents toward it for feeding.

Found from Greenland to the Florida Keys.

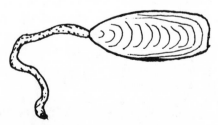

LAMP SHELL   *Glottidea audebarti* (*Broderip*)

Average one inch long.

Shell white with green lines.

Somewhat oval in shape with a straight front and a beaked rear.

Found between the tides on the southeastern coast.

Chordata and Tunicates

Chordata is a division of water breathing creatures which includes the entire group of animals that possess a cylindrical rod lying dorsally beneath a heavy nerve chord. This is a notochord which in air breathing creatures, is divided into sections called vertebra, and which in them encloses the great dorsal nerve. They have gill slits.

The ascidians or tunicates are a degenerate group of chordata that start life free swimming, with a notochord, a dorsal nervous system and gill slits, but as they mature they cease to be free swimming and settle down to a completely sessile life and lose much that they began with.

AMPHIOXIS or LANCELET

Branchiostoma virginiae (*Hubbs*)

Average length one and one-half inches.
Silvery as a fish.

This fish-like chordate is considered the in between form that links the invertebrates with the vertebrates. It has a notochord all the way down its dorsal side, a long dorsal nerve, and banded muscles. They are found in clean sand buried to their fringed hood, but when the tide is out small holes in the sand indicate their burrows. They are so swift of movement that, when dug out, they burrow so rapidly into the sand that it is almost impossible to capture them.

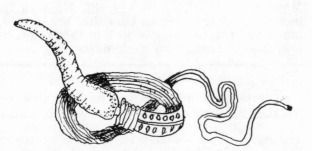

ACORN WORM *Dolichoglossus kowalevskyi* (*A. Agassiz*)

Length six inches.
Pinkish gray or yellow with slightly orange tone on body.

Acorn worms are best known by their castings that pile up in continuous coiled columns on the sand flats. They are difficult to capture. As a rule one gets only a small portion of the forward part of the creature and it looks much like an empty sleeve. They inhabit burrows and being unable to close their mouths sand flows through them forming the well known castings.

Found on the Atlantic and Gulf coasts.

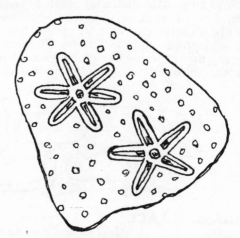

The beautiful colony of tunicates, the star tunicates, form small gelatinous crusts on gorgonians and eel grasses. They have a sexual beginning, but form their star pattern by budding.

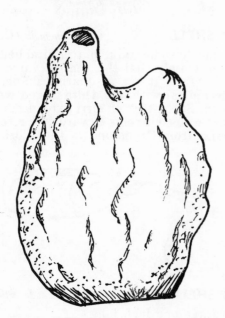

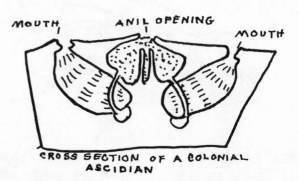

CROSS SECTION OF A COLONIAL ASCIDIAN

The large familiar sea squirt that grows in clumps or singly on pilings and sea walls squirt dirty water at the slightest annoyance and so have earned the name of sea squirt.

Great masses of hard gelatinous colonies often cast ashore are frequently mistaken for ambergris. A close look will reveal the small animals within this heavy tunic.

Hippocampus

SEA HORSES

Sea horses have been a delight and curiosity to man since the time of the ancient Romans. Pliny, the Elder, (A.D. 23-79) gave them their name — "Hippus" meaning horse and "campus" meaning caterpillar. Because they are such strange creatures, a lot of strange and erroneous ideas surround them.

Sea Horses are thought to possess excellent medicinal properties, although no serious researcher has ever accepted this line of thought. They have been ground up and used in love potions because of their supposed aphrodisiac properties. Mixed with oil of roses, they have been used as a cure for chills and fevers. The poor creatures have even been prescribed as a cure for that curse of the middle-aged man — baldness.

Sex Life

Fortunately for the sea horse he is a relatively prolific breeder. Even this basic and essential ritual is most curious. The male is the one that gets pregnant and gives birth. The female deposits her eggs via a phallus-like oviduct into a pouch in the underbelly of the male. Depending on the species, the eggs mature in about ten days. Two days after delivery, the male is ready to mate again. The offspring grow rapidly and mature within two to three months.

According to extensive studies by Dr. Kirk Strawn, the Florida Pigmy Sea Horse breeds nine months out of the year, November through January being the unproductive months, probably due to the cold water. Two broods of sea horses a month are normal when the water temperature runs 85 degrees Fahrenheit. Eggs average 25 per clutch for the Pigmy Sea Horse but as many as 55 have been noted. Specimens of the larger Lined Sea Horse found in Florida's Biscayne Bay had as many as 250 to 600 eggs in the male's brood pouch.

Where Found

Sea Horses are generally found in shallow bays where the bottom is covered with grass. They have a prehensile tail that they use to grasp a blade of grass or other marine growth. They are very slow swimmers, one writer states that it would take five minutes for one to cross a bathtub. This lack of speed should make them easy prey for predators except that their efficient protective coloring and slow motion allows them to blend with a seaweed background so effectively that they are seldom seen underwater.

In Florida, sea horses are found wherever there are shallow, grassy bays. At Cedar Keys, Tarpon Springs, Tampa Bay, and on down the West Coast of Florida, they are fairly abundant.

Vanishing Sea Horses

The siltation caused by dredging in Florida bays is a direct threat to their existence. When a dredge moves into a bay, the larger sand grains are pumped onto a fill and the smaller silt particles run back into the bays as sludge. The grassy flats are destroyed forever where the dredge has actually dug. This is not the end of it, it gets worse. The silty sludge migrates out beyond the dredge's working area and smothers whatever remaining grass is left. Each time a storm comes up, the clay-like silt particles are stirred up from the bottom of the bay to cloud additional areas. The water becomes opaque and light cannot penetrate it. Where there is no light, there is no grass. Where there is no grass, there are no sea horses — or for that matter, any other marine life.

How Many Species?

There are at least two dozen species of sea horses. They are scattered throughout the world and range in color from red to bright yellow. Six are known on the American Atlantic Coast and two have been identified on our Pacific Coast. Three species have been collected from the Mediterranean and the Atlantic Coast of Africa. The Indo-Australian seas have produced ten known species.

Aquarium Pets

Sea Horses are popular salt water aquarium specimens, but like any salt water fish, they are difficult to keep. Some species do better than others. The Florida Pigmy Sea Horse has proved to be the most hardy. The Atlantic Sea Horse (Hippocampus hudsonius) and the Pacific Yellow Sea Horse (Hippocampus kuda) are also popular. These two seldom breed in captivity, whereas generations of the smaller Florida Pigmy have been successfully raised. Sea horses aren't known for their longevity, nineteen months is the longest one has been kept in captivity. Two to three years is the best estimate of their maximum life.

What Do They Eat?

Since sea horses feed on planktonic swimming organisms, they require living food to be kept in captivity. Aquarists usually hatch baby brine shrimp for their sea horse pets. Young sea horses have been known to eat 3600 brine shrimp during a 10-hour feeding period. They have developed a powerful suction that they use with pin-point accuracy. Since they cannot chase their prey, they suck in small organisms swimming within a one and a half inch range of their snout.

Hippocampus

SEA HORSES

Mediterranean
Sea Horse
(Hippocampus guttulatus)

Yellow
Sea Horse
(Hippocampus kuda)

Florida Pigmy
Sea Horse
(Hippocampus zosterae)

Atlantic American
Sea Horse
(Hippocampus hudsonius)

YELLOW SEA HORSE   *Hippocampus kuda*

A Pacific Ocean species that is readily identified by its yellow color and speckled brown spots. Found in tropical waters of the Indo-Pacific.

FLORIDA PIGMY SEA HORSE

Hippocampus zosterae

The accepted common name for this diminutive species is the Dwarf Sea Horse. Does well in captivity, grows to 1½ inches. Found in Florida from Biscayne Bay to Pensacola. Common in Cuba, Bermuda, and the Bahamas. Not long-lived, few live to be one year old.

MEDITERRANEAN SEA HORSE

Hippocampus guttulatus

Easily identified by the long spike-like projections on his head and back. The large dorsal fin is another key identification feature.

ATLANTIC AMERICAN SEA HORSE

Hippocampus hudsonius

One of the larger species, grows to 5½ inches in 10 months. Relatively hardy, has been raised in captivity. Like other sea horses, has regenerative powers and can grow a complete new fin in two weeks. Color varies from white, dark blue, and brilliant red.

Index

Index